Management of large capital projects

Management of large capital projects

Proceedings of the conference held
in London, 17–18 May, 1978

The Institution of Civil Engineers, London, 1978

ORGANIZING COMMITTEE:

M. Milne (Chairman)
R.J. Bridle
A.V. Hooker
D.S. Lawrenson
T.F. Lederer
R.W. Postlethwaite
H.O. Smith
R.L. Wilson

PRODUCTION EDITOR: Bonny J. Harding

ISBN: 0 7277 0066 9

Published by the Institution of Civil Engineers, and produced and distributed by Thomas Telford Ltd, PO Box 101, 26–34 Old Street, London EC1P 1JH

Typeset by MHL Typesetting Ltd, Coventry

Contents

Opening address

Sir Peter Carey, KCB*

Britain has had many successes in the field of its contracting industry in overseas business, but international business continues to expand. This is not a time when we can sit back and rest on our laurels. We have done well. Certainly we can do better still. In support of contractors the Government has set up the Overseas Project Board. Its function is to help with any serious difficulties which may be encountered in pursuing overseas business. Of course the Export Credits Guarantee Department is still playing an increasingly vital role.

No-one should be in the slightest doubt about the importance of large capital projects to this country: not only abroad, in overseas earnings and in spreading the reputation of British skills, but also at home, in the generation of wealth, the creation of employment, the production of services to the public, and the encouragement of inward investment.

The Department of Industry has had a strong interest for many years in project activity in this country. We have an open-door policy towards inward investment. We have derived great benefits from it and I believe that they will continue in the future.

The standard of performance on major project sites in Britain has been poor. This is recorded in the Meadow Report on Large Sites of 1970[1] and, more recently, in the Mechanical and Electrical Engineering Construction EDC Report in 1976.[2] It is a source of major concern when we are trying to induce overseas firms to establish large capital projects in this country. It is hoped that progress can be made through the Economic Development Committee towards more broadly acceptable conditions of service

---

*Permanent Secretary, Department of Industry

and basic rates of pay, the elimination of many causes of disputes, and better understanding between all parties — clients, contractors and work-force — leading ultimately to a much improved performance. This is of great and fundamental importance where large contract business is concerned in this country. Britain's completion times, when compared with those in Japan and Germany, are not good enough and a major improvement is necessary.

The principal concern of the Department of Industry is with manufacturing industry as a whole. Its objective is to improve competitiveness, in order to increase market share both in the domestic market and in export markets abroad. The sector analysis going on in the Government's industrial strategy sector working parties has thrown up weaknesses. Now companies need to analyse these and find cures for them.

This calls for a great deal of effort — tact, imagination, determination — by all those concerned. An important part of this is the manufacturing unit which comprises a number of staff with specialized functions: production planning, procurement of materials, components and subassemblies, design, production control, machining and fitting, packaging and delivery, marketing, and a number of other duties less directly affecting day to day manufacturing processes, such as financial control, employment and training, and investment planning.

The aim of the manufacturing unit as a whole is to produce the right product at the right price at the right time. Many of the individual functions are directly associated with that aim. The right product is achieved because marketing information has been used to guide the production planning of design staff, and there has been sound production by the machinists and fitters.

The right price comes from efficiency, high quality machining and finishing, thanks to good design and efficient production control, low cost packaging and good delivery. The manufacturer's goods will arrive at the right time if there is efficient control, efficient work by machinists and good delivery.

None of these functions could stand individually. There are always several which are closely associated in any productive activity, particularly in major project contracts, where a variety of successive activities are strongly linked together in time, their progress being programmed by critical path networks.

There is extensive interdependence between all parts of an

enterprise, the responsibilities to the others being discharged in two ways: by full co-operation in the co-ordinating functions of management, and by excellence in each part's own performance. If performance falls to an unacceptable level in any one area, the success of the whole enterprise may be imperilled. Fully integrated team-work is of critical importance in all areas. Because large capital projects frequently require contractors to bring resources together from several sources, the establishment of that team-work and continuing effort in it by all who participate can be far from easy, especially where local labour has to be used. However, it must be achieved if the best prospects of success are to be secured. That theme of team-work underlies the whole of the Government's industrial strategy.

We shall not succeed in revitalizing manufacturing industry in this country unless we realize that all the parties involved — government, management and labour — have to work together with a common purpose.

References

1 NATIONAL ECONOMIC DEVELOPMENT OFFICE. *Large industrial sites*. Report of the working party on large industrial construction sites. HMSO, London, 1970
2 ECONOMIC DEVELOPMENT COMMITTEE. *Engineering construction performance*. Report of the comparative performance working party: mechanical and electrical engineering construction. NEDO, 1976

1. International sources of finance for large capital projects

S.M. Yassukovich*

Introduction

1. International finance in the last two or three years has been dominated by so-called sovereign risk borrowings by both the industrialized countries and the developing countries. International bankers have found themselves assuming a role within the international monetary system which used to be confined to the major international development agencies, the central banks, in short the official world. Traditionally international banking is concerned with the finance of trade and investment across international borders. Many bankers feel extremely uncomfortable with the present state of affairs. They find themselves forced to make credit judgments based on political assessments, and then become susceptible to political pressures of a new variety which they are not always equipped to cope with.

2. This new situation was precipitated by the four-fold increase in the price of oil in 1973 which, in turn, led to a major disruption in the world balance of payments structure. In the absence of adequate international monetary arrangements the financing of the apparently chronic balance of payments deficits of both industrialized and developing oil importing countries has been left largely to the private banking system. Although this system has responded adequately and indeed has confounded the sceptics by its ability to handle the mammoth task, there is growing concern that the concentration on balance of payments financing has crowded out the needs of commercial and industrial development projects around the world.

3. In order to put the volumes in perspective, in 1977 some US

*Managing Director, European Banking Company Limited

$ 31 000 million of syndicated medium-term eurocurrency bank credits were recorded, US $ 11.5 billion was borrowed by countries covered by the Organisation for Economic Co-operation and Development, US $ 10 000 million by developing countries. Somewhat paradoxically US $ 6.6 million was borrowed by oil exporting countries which are financing current development by indirectly pledging future oil reserves. Less than 20% of total borrowings in the banking market represented corporate borrowing and the bulk of the remainder represented balance of payments financing with no specific allocation of use of proceeds. In addition the international bond market produced a further US $ 30 000 million of financings. In this case US $ 23 000 million of this total represented OECD country borrowings; private corporations had a larger share of funds from this market, just under 50%. The rest was borrowed by governments, public agencies and international organizations.

4. The concern of the banking community relates largely to its inability to correctly monitor the use of funds which it lends to governments for balance of payments purposes. The International Monetary Fund, as we know from our own experiences in this country, is able to establish precise conditions and financial targets and therefore politically sensitive criteria, when negotiating assistance to a member country in balance of payments difficulties. It sends teams to monitor adherence to these conditions and criteria. The private banking system has no such political power, no sanctions, except the withdrawal of future facilities, which can merely endanger the loans already granted. There is always a sensitive balance of power between lender and borrower; some observers feel that in the international capital market the balance of power has shifted towards the borrower in a number of key countries and the banking system may be forced to provide almost permanent facilities in the hopes of eventually recovering its original loans.

5. The excessive concentration of international banking assets in balance of payments financing has focussed more effort on developing techniques for the financing of specific projects. The term 'project financing', like many expressions in universal use, is somewhat misleading. It has come to mean, in its broadest sense, finance directed at a specific capital investment in contrast to balance of payments financing, the purpose of which is to increase the international reserves of a country. Rather than merely describing current

sources of international financing in a general sense, this Paper concentrates on the methods used to evaluate the risks associated with major capital projects around the world, the techniques employed by bankers in devising financial packages and the sources of finance which are incorporated in these packages.

6. In order to describe more precisely techniques of risk assessment and financial packaging in the field of project related lending and financing, the expression project finance will be treated in its purest form. However, it is important to understand that very few capital projects in the international sector are susceptible to the pure project finance approach. There will almost always be a mix of public or private sector support in the form of guarantees and other means of recourse together with cash flows from the project itself. Financing will therefore tend to be supported by a combination of guarantees, security and revenue which will vary from case to case.

7. It is generally accepted that project financing is defined as the financing of a particular economic unit, wherein the lender looks solely to the present and future cash flows and earnings of the economic unit as the source of funds from which debt will be serviced, and to the assets of the economic unit as collateral for the debt.

Basic characteristics of project financing

8. Most project loans will usually exhibit several or all of the following characteristics.

(a) Project financing techniques are more readily applicable to funding the extraction, processing and transportation of a natural resource such as hydrocarbons, minerals and timber, or to financing the construction of a pipeline, refinery, or petrochemical facilities.

(b) The project sponsor may include one or more shareholders or shareholder groups (companies and/or governments).

(c) A substantial amount of financing is required.

(d) Although it may be a large expansion of an existing facility, the project is usually a start-up operation.

(e) Firm purchase contracts are provided for a term satisfactory to the lenders.

(f) Net cash flow of the project is expected to service all of the associated debt financing.

Risks

9. Often the project sponsor(s) may prefer, or be required, to obtain financing based solely on the economic viability of the project in order to minimize or avoid the consequential impact to their credit standing or balance sheet. This type of loan would be classified as being completely non-recourse to the borrower. Since prudent lenders must be assured that they will be repaid by either the project, the sponsor or a third party, it becomes extremely important to make a thorough assessment of the various project risks. The magnitude of each risk must be defined and deemed to be acceptable to both the borrower and lender before a project financing can be successfully structured and completed.

10. The most difficult hurdle to overcome in project financing lies in structuring the loan with a minimal amount of recourse to the sponsor, while simultaneously providing sufficient credit support, through guarantees or certain other types of undertakings of the sponsor or third party, to assure the lenders that the credit risks are satisfactory and acceptable. The most commonly held misconception regarding project financing is that this technique is always 'non-recourse' in nature and 'off the balance sheet' of the borrower. In an effort to avoid misunderstandings and disappointments which may result from such misconceptions, the lenders must carefully advise the prospective borrower as to which projects can be financed by this technique and specifically how.

11. Only a careful assessment of the risks by the lender and borrower can prevent this embarrassing and often costly event from occurring. The eight principal risks which are common to many project loans and against which lenders will seek protection are as follows:

(a) credit risk of project sponsor
(b) natural resource or raw material
(c) design and construction
(d) operating and management
(e) transportation

(f) market or commercial
(g) political
(h) *force majeure*

Discussion of project risks

Credit risk of project sponsor

12. It is very important for the lender to be completely satisfied with the credit-worthiness of the project sponsor (companies and/ or governments). Even though project financing may be satisfactorily arranged, it will usually be necessary to require certain undertakings or financial guarantees initially (perhaps during the construction period), until the project is fully completed and operational. Unexpected cost overruns may be incurred before project completion which may require additional equity and/or financing provided by the sponsor. The sponsor must have the financial ability and reputable and financially sophisticated management able to cope responsibly with such contingencies.

Natural resource or raw material risk

13. Lenders must be satisfied that the natural resource being developed exists in sufficient quantity and quality, or that raw materials or feedstocks are contractually committed (both in amount and within a specific time frame), to enable the project to meet its commitments and to achieve comfortably its cash flow projections. Lenders frequently require confirmation of the natural resource reserves and production capability by an independent and reputable consultant. The cost of such an evaluation would be for the account of the project or its sponsors.

Design and construction risk

14. It is not uncommon for project sponsors to include in any project financing a request for sufficient funds to construct the required project facilities. Since it is normally not possible to determine exactly what the final construction cost will be until

9

project completion, and as the servicing of any debt obligations relating to a previously arranged project loan will be fully dependent on the net cash flows generated only by a successfully, continuously operational project, the lenders will require additional support from the project sponsors. This undertaking from the sponsor will normally be provided in the form of a completion guarantee. Although some borrowers are reluctant to provide completion guarantees of any variety, lenders believe that if this type of undertaking is not given they may be taking an unacceptable equity risk in lending to the project. Often a project loan is structured in such a way that it only becomes non-recourse after final completion, thereby relieving the lender of this one risk and enabling the borrower to re-finance his project.

15. Usually a completion guarantee required by lenders from the project sponsors will be structured in such a way as to unconditionally guarantee that even if the original capital cost of the project is exceeded, the project facilities will be completed by a specific date (certain time delays may be allowed), and will operate at the designed capacity and at a continuous production level to produce the quantity and quality of the product required. If the project being financed incorporates new technology of some type, lenders may require an unconditional financial guarantee from the project sponsors until the project completion date and until the new technology has been proven by successful and continuous operations. The lenders would then allow the guarantees to lapse.

16. Even though project lenders employ experienced specialists who easily understand construction, design and new technology, it may be necessary to require independent confirmation of the technical and economic feasibility of a project by a reputable consultant. The cost of the evaluation would be for the account of the project or its sponsors.

Operating and management risk

17. The following important questions must be satisfactorily answered and carefully evaluated in reviewing the operating risks.

(a) Can the operator demonstrate availability of management

with proven ability and employ a sufficient number of qualified and experienced personnel?
(b) Does the operator have previous experience with the complexities of extraction, processing or transportation technology used in the project?
(c) Does the location of the project cause any unusual concern regarding remoteness, climate, work-force and infrastructure availability?
(d) Are there any environmental concerns regarding water/air pollution and reclamation?
(e) Does the operator exhibit the ability to estimate and maintain predictability and variability of project operating expenses?

Transportation risk

18. If the product being produced by the project facilities must be transported to market before it can be sold, it is necessary to determine how it will be transported, and what quantity, physical state, distance and special technology if any is to be used. The appropriate transportation facility must be available and operational at the time production begins. Therefore the lender must consider this risk and even request independent consultant assistance, at the project or sponsor's expense, to confirm that proper transportation facilities will be available.

Market or commercial risk

19. The project lender will carefully review all contracts and agreements prior to structuring a project financing, particularly with regard to the marketing of the product being produced. If the lenders are looking for security offered by long term sales contracts, it is normal for certain specific provisions to be included such as the term of contract, quantity of product to be sold, reputation and credit-worthiness of buyer, pricing agreement including special escalation provisions to protect the seller against currency fluctuations and inflation, special royalty or tax covenants and *force majeure* obligations. Lenders will normally encourage the sponsor to strengthen the sales contract by offering equity in the project to the buyer, or by obtaining a financial commitment

from the buyer such as an advance payment or similar type of loan facility.

20. Financing on a project basis becomes easier if the product being sold is and will be in strong demand during the term of the sales contract. Strong product demand certainly facilitates the contract negotiations for the seller in securing firm sales contracts from the purchasers, including price escalations and other important covenants, thereby making the project more viable and successful.

Political risk

21. Lenders carefully review this risk on a country by country basis and focus on the following aspects in making their assessment.

(a) Political stability of the government or governments involved in the project
(b) Financial conditions and credit-worthiness of the country or countries involved in the project
(c) Type of project and local government's ranking of the project priority
(d) Equity participation or any other financial undertaking provided by government or governments
(e) Will the construction and operation of the project employ a substantial number of nationals?
(f) Will a government be involved in project operations and management? If so, is there a sufficient number of experienced people available to fulfil this role?
(g) Are government regulations and tax laws unduly restrictive regarding the initiation and continuing viability of the project?
(l) Stability of currency or currencies involved and convertibility
(i) Is the country dependent on only one industry or one saleable product for foreign source income and are the buyers diversified in number and location?

22. Since the political risk increases simultaneously with the term of loan repayment, lenders may often request Overseas Private Investment Corporation (OPIC) and/or commercial insurance coverage where applicable and available, which will protect the

borrower and lender regarding expropriation, war risk and currency convertibility. Recent project financings have exhibited examples of multinational ownership and financing with the intent of minimizing or reducing the political risk. Another technique used to reduce this risk is to allow a trustee independent of the seller and buyer countries to act as the administrator of drawdown and repayment of all associated project loans and excess funds.

Force majeure risk

23. Unusually lengthy delays in project completion and/or interruptions in continued operations after start up will increase project development costs and cause difficulties in servicing any debt obligations incurred by the project sponsor. Lenders will therefore want to review any *force majeure* provisions included in agreements relating to the project to be financed, and be assured that no unreasonable penalties or other financial obligations conflict with the proposed financing.

Debt:equity ratio

24. One of the questions most frequently asked by prospective borrowers contemplating project financing is how much equity is required to support the proposed financing. There is no precise answer that can be given or standard guidelines to follow because of the many variables that must be analysed by the lenders before a debt:equity ratio can be established. A number of project loans have been successfully structured and completed with equity amounts varying from 0% to 60% of the total cost of the project. In other words, the amount of debt that can be raised on a project basis is a function of the lender's assessment of the various risks involved, validity and achievability of estimated cash flow forecasts, and the recognition of specific limitations regarding amount and term governed by the economic environment existing at the time and financial sources available, which may be utilized to finance the project.

25. In general a greater amount of gearing or leverage can be achieved on a project basis than for a typical manufacturing and industrial company because cash flow projections can be monitored,

controlled and made more predictable, and a number of risks can be minimized or reduced, by establishing firm take or pay contracts, throughput and deficiency agreements or similar type agreements. The amount of gearing will also depend on the aims of the project sponsors (such as certain anticipated equity returns) and the amount of debt the project will support.

Possible financial sources available for project financing

26. Most sponsors investigating the possibility of financing their development on a project basis will try to minimize the possible currency risks encountered in the future by borrowing or raising the maximum amount of debt in a similar currency to that in which the project will generate revenues.

Buyer credit

27. Since the purchaser of the product being sold by the project may be interested in obtaining a secure supply at a reduced price, and specific minimum amount of the product on a long term basis, it is possible that credit and/or financing may be available either in the form of an advance payment, or subordinated or unsecured loans for a portion of, or the entire term of, the sales contract from the buyer.

Local financing

28. If financing is required in the currency of the country where the project is being developed, the project has local ownership, and the government of the country has rated the project as top priority by providing financial support or certain tax or other attractive incentives, then local bank, insurance and other term lenders may be willing to provide a certain amount of attractively priced and longer term financing to the project sponsor.

Export credit agencies

29. A number of export credit agencies of governments in

developed countries offer very attractive financing, both in terms and interest rates, enabling a project sponsor to obtain equipment and certain engineering services in substantial amounts. These export agencies are very competitive and it therefore behoves the borrower to carefully investigate and compare the terms and conditions offered before selecting the appropriate financing. Proper co-ordination of these credit alternatives from a number of countries must be monitored and directed by the lenders. The agency will guarantee the financing provided by commercial banks (usually with an incremental interest rate subsidy) or provide the financing itself.

Government guarantees

30. It is possible in certain instances to obtain load guarantees from a government or government development bank to support a project development in their respective country. This type of undertaking will encourage other lenders to participate in the loan and enable the necessary amounts to be raised.

Eurocurrency bank financing

31. During the last few years, this funding source has enabled banks to raise substantial amounts for project loans. Most of the major project loans have utilized this source of funding by first selecting a managing bank experienced in project lending to structure the appropriate financing required, and then requesting a group of banks experienced and interested in financings of this type to participate in the loan, while simultaneously placing particular emphasis on type of project, geographical location, country interest and well established sponsor relationships.

Eurocurrency bond market

32. The Eurobond is usually confined to standard corporate or government obligations because of the lack of interest by investors in start-up or developing projects (pipelines are the exception). This source may, however, be tapped later to re-finance project debt.

Commercial bank financing

33. Although commercial bank financing is available in substantial amounts for project financings of various types, such loans are medium term in duration and may not coincide with the project requirements, particularly mining developments. Usually this lending policy will limit the bank to financing construction and only a portion of the term financing. However, certain types of natural resource development, such as oil and gas, will exhibit more rapid payout periods, and banks may be able to finance the entire project loan.

Long term financing

34. A number of projects often require longer term financing than commercial banks are willing to offer. The public debt market is not usually accessible to finance projects of any type and it is, therefore, possible in certain situations to utilize the private placement market for longer term fixed rate financing. Banks and long term lenders, such as insurance companies, have well established working relationships that make it possible and advantageous for one financial institution to provide a project financing in total.

Conclusion

35. The following is an outline of a number of characteristics that most viable and successful project financings have in common.

(a) Every project financing must have some form of strong credit support either as direct or indirect guarantees, take or pay contracts or economic necessity provided by the sponsor or by a third party. It is possible to structure these various types of undertakings in such a manner as to minimize or avoid stating them as straight debt on the balance sheet.

(b) Lenders are willing to accept a project credit risk but not an equity risk. Bankers lending to projects or on any other basis are definitely not in the business of taking equity risks. There are recent examples where lenders have been specially compensated for taking equity risks but it is believed to be an exception rather than a general rule.

(c) Before any financing can be arranged, appropriate feasibility and engineering studies must be prepared to confirm conservative cash flow projections and the viability of the project. Sufficient cash flow must be available in the future to service all debt, working capital, operating costs and also provide an adequate coverage ratio or cushion (cash flow/debt) for any contingencies that may arise such as cost overruns.

(d) Completion of the project and the fact that it will operate in accordance with design, production and cost specifications must be assured. The project sponsor must exhibit the technical expertise and the financial capability to initiate, complete and operate the project successfully.

(e) Proved and reliable technology that has been thoroughly tested over the years in a number of other projects is usually involved. If new technology is being utilized the lenders must recognize that more than a credit risk must be evaluated and considered.

(f) The project sponsor or designated operator must provide the required qualified and experienced management and personnel to operate the project facilities in a prudent and professional manner.

(g) The collateral value of the fully developed and operating property and facilities being financed must be sufficient to satisfy the lenders.

36. It is extremely important in very large projects, for the manager of the proposed financing to be knowledgeable and experienced enough to comprehend what terms are acceptable to a group of sophisticated international lenders (long-medium-short term) and yet compatible with the terms most advantageous to the project sponsors. It is important for the project sponsors to select a financial adviser early enough to allow sufficient time to appropriately structure and complete a financial plan regarding the entire project development before approaching the lending syndicate. By utilizing this technique, delays and additional negotiations may be avoided which normally cause increased costs and may prevent taking advantage of attractive and available funding sources. The financial adviser and manager can be separate and independent financial institutions.

2. Export Credits Guarantee Department facilities for overseas capital projects

K. Taylor, CB *

Introduction

1. The Export Credits Guarantee Department (ECGD) has been in the business of giving support to British exporters for nearly 60 years. During that time trading conditions have changed considerably and ECGD facilities have been adapted and developed to meet them. One of the most significant changes has been the swing from the simple supply of goods to foreign buyers — financed by the provision of credit by the supplier, or from third parties introduced to the buyer by the supplier — to the concept of package deals under which British firms are responsible for the design, supply, construction/installation and commissioning of complete projects, similarly financed as a package. This approach involves not only commercial and technical responsibility but also financial liability for the contractor.

2. Credit insurance for exports of consumer and the simpler types of engineering goods still represents the biggest part of ECGD's business; of the £12 000 million of exports insured in the year to March 1977, some four-fifths involved credit of six months or less. However, the growth in importance to the UK economy of 'project' business and recognition of the large potential market for such business — particularly in the Middle East — has been reflected in a considerable concentration of ECGD's resources on its support. This is illustrated by the fact that, of the 15 divisions making up ECGD's organization, seven deal largely, if not wholly, with project-related business, and much of the work of the Department's 'common services' divisions, such as those concerned with country assessment and international relations, is devoted to the special

---

* Secretary, Export Credits Guarantee Department

needs and problems of these areas of business.

3. Virtually all UK exports sold on credit terms of two years or longer are insured by the Department. Although exporters may, where projects are to be paid for on cash or near-cash terms, be willing to take on the business without ECGD cover against non-payment, few complete projects involving credit terms are undertaken abroad by British companies without the benefit of ECGD basic insurance facilities. Also, changes in world market conditions, particularly those stemming from oil price increases, have led to the need for new forms of support. ECGD has responded by extending the range of its facilities in a number of ways, mostly in order to meet the special circumstances affecting exports of the heavier types of capital goods and the involvement of British contractors in major projects.

4. This Paper attempts a summary of ECGD's principal relevant facilities, which those engaged in projects business may want to draw on in various permutations according to the character of their own business and of the overseas project concerned. It is particularly important that exporters of this class should contact ECGD as early as possible once a potential project is taking shape, so that the best permutation can be worked out early and so that, since ECGD has to operate on a commercial basis, the exporter can get best value for his premium payments. At the even earlier stages of identifying potential projects and mobilizing the British effort, the Department of Trade may be able to help, and a summary note about its Overseas Project Group is in the Appendix at the end of this Paper.

Credit insurance

5. ECGD's traditional role is that of credit insurer, i.e. giving cover against a range of possible causes of non-receipt of payment by UK exporters from overseas buyers of their goods and services. This 'Supplier Credit' insurance is provided through various forms of guarantee tailored, where necessary, to meet the needs of particular trades. Exports of consumer and engineering goods, where business is of a continuous and repetitive nature, are covered by the Comprehensive Short Term Guarantee and, where sold on credit in excess of six months, by the Supplemental Extended

Terms Guarantee. For capital goods sold on a Supplier Credit basis, the Department's Specific Guarantee facility is available.

6. Companies in the constructional works field have policies specially tailored to their requirements — the Department's Constructional Works Guarantee provides protection against the non-payment of sums due for work performed. The special position of the Construction Industry is fully recognized, and the Department is in regular contact with the Export Group for the Constructional Industries to ensure that its changing needs are reflected in the support available from ECGD.

7. Cover is also available in respect of payments due for services rendered by British companies to their overseas clients, e.g. the provision of professional and technical advice. 'Services' cover is provided by guarantees similar to those available for the supply of goods. Where the business is of a continuing nature, a Comprehensive Guarantee is available; where, as in consultancy contracts related to project business, services are longer term and individual in nature, Specific Guarantees are available.

Export credit finance

8. The Department's Supplier Credit insurance cover can be supplemented by direct guarantees to banks, to assist in the generation of finance for the insured business by offering unconditional 100% guarantees to banks as security. For exports on less than two years' credit the interest rate is a floating one: up to $\frac{5}{8}\%$ over the London Clearing Banks' Syndicated Base Rate. For Supplier Credit contracts involving credit of two years or more, finance is facilitated at an interest rate which is fixed for the life of the contract and is set at a level designed to match the terms available from competitor countries. Bank guarantees are usually operative when the goods have been shipped, the works completed or the services performed. However they can generate finance earlier if the buyer is prepared to contract on a basis that involves his acceptance of a firm obligation to honour bills of exchange or promissory notes issued before those points in time.

9. In some circumstances — particularly where large amounts and long credit are involved — the use of the Buyer Credit technique is more appropriate. This technique involves a UK bank or syndi-

cate of banks providing a loan, whose repayment is guaranteed by ECGD, direct to the overseas buyer or to a borrower acting on his behalf, thus enabling the UK contractor to enjoy the benefits of a cash contract and to arrange more easily for progress payments during production. In many cases the loan is related to a single contract, but can also be made available to buyers to purchase their requirements for the fulfilment of a project from a number of different UK suppliers. Buyer Credit guarantees are available for contracts of £1 million or more involving credit of two years or more. Finance is provided on the same fixed preferential basis as with Supplier Credit bank guarantees, for up to 85% of the contract value, with the buyer being required to pay the balance direct to the supplier from his own resources, including an adequate payment on signature of contract.

10. To support these facilities for fixed preferential rates of interest, ECGD has arrangements with UK banks for offsetting differences between these rates and an agreed commercial rate of interest. In the case of sterling credits, there are further arrangements whereby ECGD re-finances a part of the banks' commitments; hitherto, the fixed-rate sterling facility has in practice only been available from the clearing banks but as from 1 April, 1978, all banks which are authorized by the Bank of England under the Exchange Control Act 1947 and which are incorporated as companies in the UK will be eligible to participate. In the case of credits financed in approved foreign currencies, there is no automatic re-financing arrangement, but ECGD is prepared to take over any part of a loan which the banks cannot continue to fund on acceptable terms. This currency finance scheme is open to all banks defined as above, with the addition that non-incorporated branches in the UK may participate in syndicates provided that the syndicate is not led by a foreign bank owned by the same group.

11. In providing this support for both Supplier Credit and Buyer Credit transactions, ECGD operates within internationally agreed guidelines relating to minimum downpayments, minimum interest rates and maximum lengths of credit. These minima and maxima vary according to the financial strength of the buyer's country, and there is provision for the matching of any more favourable terms that may be offered by official credit insurers in other countries, thus maintaining the UK's ability to compete on credit terms.

Foreign currency arrangements

12. Where exporters contract in foreign currencies and either use the forward exchange market to protect themselves from losses resulting from exchange rate movements, or use foreign borrowing to finance their contract, the various forms of credit insurance facilities provided by ECGD can be extended to cover the situation that an exporter would face if, having failed to receive payment from his customer, he had to purchase foreign currency, at a cost greater than the Department's sterling claims payment, to enable him to meet his forward exchange sale or loan repayment obligation. For business covered by Comprehensive Short Term Guarantees, ECGD's claims payment can be increased by up to 10% in such an event. For business covered by Extended Terms or Specific Guarantees, the exporter can select any additional margin of cover he wishes, against a *pro rata* increase in his premium and, in the case of contracts financed in foreign currency on a Buyer Credit basis, ECGD's guarantee covers the full amount of the foreign currency loan.

13. For balance of payments reasons and because of the burden on the Exchequer of providing re-financing facilities for ECGD-insured sterling export finance provided by the banks, a substantial proportion of Buyer Credit business has, since February 1977, been required to be financed in approved foreign currencies, normally US dollars or DM.

14. Companies bidding for contracts expressed in foreign currencies can be faced with an unfavourable change in sterling's parity with the contract's currency after submitting their bid and before the contract award, which is the earliest point at which they could confidently cover themselves by selling forward. To meet this problem, ECGD introduced in August 1977 a facility, called Tender to Contract Cover, which guarantees UK bidders the same sterling out-turn as would have resulted had they sold forward their prospective currency receipts at date of tender. This facility is available for contracts payable in an approved foreign currency for which the exporter intends to take out a Supplier Credit or Buyer Credit cover expressed in that currency.

Performance bonds

15. Recent years have seen an increasing tendency for overseas

buyers, particularly in the oil-rich Middle East markets where a large proportion of project opportunities arise, to insist on bonds from acceptable third parties, to guarantee the performance by contractors of various obligations undertaken by them in both their bids and their contracts, which may be callable 'on demand' or in specified circumstances. In response to representations from exporters, ECGD has developed a range of facilities relating to tender, advance payment, and performance guarantees and to customs, maintenance, retention and progress payment bonds.

16. Under these facilities, ECGD's willingness to issue a guarantee to a bank or surety company can encourage the issue of bonds in circumstances where it might not otherwise have been forthcoming, since the guarantee provides a 100% indemnity to the bond issuer against the consequences of the bond being called. The scheme allows recourse to the exporter only in the event of his failure to perform and, where several companies are involved jointly in a project of such size that joint responsibilities would impose excessive burdens on the participants, ECGD is prepared to consider limiting its recourse to each of the partners to the value of its individual share of the contract. When first introduced in February 1975 the Department's bond support scheme applied only to contracts valued at £20 million or more on cash or near-cash terms and where a firm had demonstrably exhausted its commercial bonding capacity. This latter condition has since been dropped and access to the scheme has been progressively lowered to cover contracts valued at £500 000 or more.

17. Where companies can raise bonds without the bond-issuer needing direct support from ECGD, they can obtain cover against the unfair calling of the bonds as an optional extra to ECGD's basic credit insurance guarantees. This cover is available for any contract credit-insured by the Department irrespective of its value or terms of payment.

Cost escalation cover

18. In 1975 ECGD introduced on a temporary basis a scheme to give exporters a degree of protection against the consequences of the high and unpredictable rates of inflation then prevailing

in the UK, which placed them at a disadvantage when bidding for large contracts with long delivery periods where the buyer was able to insist on fixed prices or on very limited commercial escalation provisions. To be eligible for this scheme, contracts must have a value of at least £2 million and a production period of two years or more; in contracts for the supply of a number of similar units, each individual unit must be worth at least £500 000. The exporter is covered against a proportion of annual increases in variable costs above a pre-set threshold and below a certain ceiling. The higher the threshold he selects, above the minimum specified in the scheme, his ceiling of cover will rise more than proportionately, and is further raised in the case of cash contracts and all contracts financed in foreign currency.

Project indemnity risks

19. The problems arising from buyers' requirements that consortium members accept joint and several responsibility to the full value of the contract for the consequences of the failure of any of them, or that main contractors should carry unlimited liability for the consequences of the failure of a sub-contractor, led to the introduction in 1975 of ECGD's Project Participants' Insolvency Cover. This scheme provides protection to consortium members, or a main contractor, against the failure of a fellow member, or of a sub-contractor, to meet his contractual obligations to the insured as a result of his insolvency. Following representations from exporters that this scheme and the facilities of the private surety market still did not adequately cater for the indemnity risks arising in very large overseas projects, it was announced in December 1977 that the ECGD scheme would be extended to cover losses which may arise in a situation falling short of the actual insolvency of a UK consortium partner or sub-contractor. This broader facility will be available for an experimental period of three years on a selective basis for overseas projects which are considered to be particularly attractive on national interest grounds and of which the value is at least £50 million. Details of the scheme are still being finalized.

Pre-shipment finance

20. The finance generated by ECGD's bank guarantee facilities normally becomes available to the exporter on shipment of goods or completion of the work involved, since foreign buyers are not always willing to agree to progress payments as the work is being done. To assist exporters who experience cash-flow problems during production, ECGD is prepared, for a contract of at least £1 million and a manufacturing period of one year or more, to provide a guarantee to a bank in respect of the working capital required for that contract during the 'pre-shipment' period. Funds generated by this scheme attract no interest subsidy or re-financing by ECGD.

Multinational projects

21. Very large projects sometimes require resources from more than one country; the co-ordinated provision of goods and services by companies in a number of countries will often benefit from similarly co-ordinated support between their credit insurance and financing institutions. ECGD recognizes the need to meet the complex problems arising out of multinational projects, and participates actively with its foreign counterparts in the development of techniques to meet these problems. Within the EEC, and on a reciprocal basis with certain other European countries, there are arrangements whereby the national export credit agency such as ECGD will extend its facilities to include a proportion of goods or services from the other countries — up to 40% of the main contract value in the case of the EEC and up to 30% in the other cases. There are as yet no standard inter-governmental arrangements for multinational contracts whose composition does not fall within these proportions, but ECGD is prepared to seek *ad hoc* forms of collaboration with other insurers where the need arises.

Appendix: the Overseas Projects Group of the Department of Trade

Functions and objective

22. The Overseas Projects Group (OPG) co-ordinates official assistance, at home and abroad, to British firms in pursuit of major contracts in large capital projects overseas. It acts as a focal point within the government machine to which British consultants, contractors and manufacturers can look for assistance, especially in complex and difficult cases. Its services are aimed at improving the net cash flow to the UK balance of payments.

23. OPG's assistance is directed towards those projects which require a package or turnkey approach. The package might include elements of consultancy, design, technology, construction, supply of plant, materials and equipment, finance and training. OPG does not become involved in contracts solely for the sale of hardware, ships or aircraft. It tries to ensure that there is only one British bid in the projects in which its assistance is required, so as to make the best possible use of the services which the Government can offer and the influences which it can bring to bear, and to avoid wasteful competition between British firms.

24. OPG operates a policy of selectivity according to size of project, industrial sector and geographical area. In general it looks for projects which offer opportunities for British exports of goods and services of at least £10 million; it is expected that the value of the goods will usually exceed the services element. OPG concentrates on the industrial sectors in which British capability is good and on geographical markets where payment presents no major problem. Many of these markets are in the richer developing countries which do not have the manpower or technical resources to cope with large and complex projects themselves, and therefore tend to seek bids for a total package.

Methods of Working

25. OPG acts on a responsive basis and will give assistance in projects which meet its criteria. Sometimes it acts on an initiative basis, for example if it has special intelligence about a new project, which it might have obtained from an inter-governmental Joint Commission. In such cases, OPG would seek, in consultation with

industry, to identify a suitable leader of a British group to tackle the project. Complex projects of a multi-disciplinary nature will frequently call for a combination of British concerns working in consortium or on a joint venture basis. In most instances British concerns are able to organize their own groupings to pursue these contracts, but if necessary OPG can assist in the formation of a suitable grouping.

Type of assistance

26. OPG's assistance falls basically into two categories: administrative and financial. It co-ordinates the assistance which government services are able to give in the often long pre-contractual period, e.g. help from the appropriate overseas post or from a government department or public sector organization with expertise for the particular project. In certain cases OPG will give financial assistance from the Overseas Projects Fund.

Overseas Projects Fund

27. The purpose of help from the Overseas Projects Fund is to encourage greater effort in pursuit of major, complex and difficult project contracts. Subject to prior agreement, OPG can contribute up to 50% of the pre-contractual expenses of companies pursuing such contracts. To qualify for assistance, the project must offer a minimum UK content of £10 million in goods and services, with the goods being the predominant element. Similar assistance can be provided for consultancy or project management contracts which by themselves would make a major contribution to the balance of payments. The minimum UK content for these contracts is £1 million in fees. The amount which can be contributed is subject to a maximum per project (in the range £50 000–£100 000) which is related to the gross value of the contract being sought, or ½% of the gross value, whichever is the lesser. In the event of the contract being won, the contribution is repayable.

Overseas Projects Board

28. OPG has the benefit of advice from the Overseas Projects Board (OPB) set up by the British Overseas Trade Board (BOTB).

The OPB is the recognized focal point for industry and Government for consultation on problems and matters of interest arising in the pursuit of overseas project business. Its membership is a mix of top level representatives of industry and senior officials in government departments. Its Chairman is an industrialist, who is a member of the BOTB. The industry representatives have wide experience covering all aspects of overseas project business. They are appointed in a personal capacity, but individual members have the responsibility of providing a close link with particular industry organizations whose members are concerned with overseas projects, so as to make the OPB representative of industry's views.

Enquiries

29. Enquiries about assistance in pursuit of a major contract overseas should be addressed as early in the life of a project as possible (provided it is beyond the preliminary feasibility stage) direct to OPG. Early application for Fund assistance is of particular importance and requests for copies of the Guidelines for the use of the Fund and necessary application forms should also be made to OPG. The Group is not the source of information in the Department of Trade on specific opportunities or market conditions. Enquiries of this kind are dealt with by the General Export Services Branch or the Commercial Relations and Exports Divisions.

3. The role of the banker in assisting the contractor to win a contract

R.J.R. Owen, MA*

1. There is a widespread feeling among British contractors that their competitors in certain other countries (France, Japan and Germany are perhaps the most frequently mentioned) somehow receive more help from their governments and/or their banks than contractors in the UK. This is not always true, particularly in relation to the role of government. However, there is one area in which the UK banks could perhaps make a greater contribution than they do at present. This is in the seeking out of projects which present opportunities for British contractors, and the whole range of services involved in actively helping a contractor or exporter to win a particular contract. In England, despite the City's pre-eminence as a financial centre, in this particular field of activity we are not as enterprising as the banks in some other countries, particularly France.

2. Many contractors have traditionally looked upon their bankers (and the bankers have looked upon themselves) as the people who are brought in to arrange the finance once a contract is won, or is at a fairly advanced stage of negotiation. To the extent that this is so, a national opportunity is being wasted. If the banker is brought in, or himself brings in the contractor, at the initial stage of a bid, his efforts can play a very important part in the winning of the order.

3. This obviously applies particularly in developing countries, where the availability of finance is often the first thing in the client's mind and the factor upon which the realization of the project depends. For this reason bankers sometimes become involved in projects even before contractors. If a banker and a contractor can jointly present a satisfactory financing proposal,

*Director, Morgan Grenfell and Co. Limited

in some countries they are already half way towards winning the
contract. If a complete package can be offered to provide finance,
supply equipment, construct and commission, it frequently
increases the chances of the project getting off the ground and of
winning the contract. It also increases the chances of obtaining a
contract on a negotiated basis. Although negotiated contracts
used to be looked upon by bankers as something slightly sinful,
the concept is beginning to gain respectability even in such places
as the International Bank for Reconstruction and Development.
Even in countries which can afford to shop around, the time (and
therefore money) saved by taking the negotiated package is frequently, in my experience, much greater than any saving through
competitive tendering.

4. In many countries, however, simply offering finance is not
enough; more often than not there is someone else offering it
also. In this situation the conditions of the finance become crucial
and the skill, energy and inventiveness of the banker come into play.
The ability to put together finance for 100% of the project cost
(including the local element) may be important, or to roll up
interest to ensure that the buyer has the minimum of payments to
make before the project is generating cash, or in some cases to
introduce an equity partner. The length of credit (both export
credit and the accompanying commercial financing, usually in
eurocurrency) and the interest rate are obvious factors. A less
obvious one, but arguably more important, is the currency — both
of contract and credit. Despite the dramatic fluctuations in currency
parities of recent years, it seems that buyers and borrowers still
generally do not take this factor into account sufficiently when
evaluating bids, and that British contractors do not receive the
advantage they should enjoy from quoting in a currency weaker
than that of most of their competitors.

5. Fortunately, the UK now has an export credit system which
allows greater flexibility in the currency of loans backed by the
Export Credits Guarantee Department than the equivalent arrangements in most other countries. The ability to use this flexibility
to maximum advantage can be a very important factor in winning
contracts and does involve a high level of banking technology.
Also important is the persuasive power of the bank in getting over
to the buyer the implications of contracting and financing in a weak
currency (such as sterling or US dollars) as opposed to a stronger

one (such as Deutschmark or yen). In situations where a British contractor is bidding in a currency other than sterling there are usually many contractual problems to be resolved in order to eliminate or minimize the currency risks being carried by the contractor. Here the previous experience of the banker in similar situations, his knowledge of the provisions of commercial contracts, his knowledge of, and ability to operate in, forward foreign exchange markets, and his ability to inter-relate all these factors, are nowadays a crucial element.

6. However, the role a banker can play in supporting the contractor goes beyond these financial factors. Through his knowledge of political and/or commercial conditions in the buyer's country, his representation in the country or area and his relationships with relevant decision-makers, he can provide important advice and information relevant to the contractor's strategy and generally support the sales effort. I have known a number of situations where this kind of help was crucial in the winning of a contract.

7. The banker can frequently also provide advice on, and assist in, the negotiation of the actual commercial contract, particularly in areas (like Eastern Europe) where extensive previous experience and knowledge of precedent is important.

Conclusion

8. The banker, particularly the merchant banker, can play an active and entrepreneurial role in support of British exporters and contracts, starting at the earliest stage of a project. Admittedly the project 'drop-out rate' for a bank operating in this way will be high. I have bitter experience of spending a large amount of time and even substantial amounts of money on projects which have been cancelled or contracts which have, despite all efforts, been lost. Indeed, as every contractor knows, it is probably necessary to put at least 10 and maybe 15 horses into the race in order to get one to the winning post. Nevertheless, this entrepreneurial approach to contract financing is something which should be encouraged. There are a number of British contractors who do try to use their banks like this, and a number of UK banks who are accustomed to operating in this way. From the national point of view, I hope this kind of activity will increase.

4. The World Bank's role in the initiation and financing of large capital projects

S.M.L. van der Meer *

Introduction

1. The World Bank plays a significant role in many large capital projects in developing countries. Such projects are typically concentrated in the fields of transportation, hydro-electric and thermal power, industry, irrigation, water supply/sewerage and telecommunications. More than half of the $7 billion the Bank and its affiliate, the International Development Association (IDA), lent during its 1977 fiscal year was for projects in these sectors and most of them fall in the category of large capital projects under any reasonable definition. Since the Bank provides only a part of the total financial requirements of the projects for which it lends, it is associated with an investment that is a multiple of the volume of its lending.

2. Its involvement in projects goes far beyond the mere provision of funds. Typically its association with a project begins at the time the project concept is being formulated and feasibility studies are yet to be undertaken. It continues throughout the subsequent phases of preparatory work, which include economic and financial analyses, detailed engineering, cost estimates and loan negotiations, and culminates in the granting of a loan or credit. At that point the necessary funds are secured and generally the project is technically sufficiently advanced to start the process of contract awards in which the Bank also has a strong interest. The Bank involvement does not end there but continues until the physical completion of the project, when an *ex-post* evaluation by the Bank jointly with the borrower takes place to establish to what extent the original objectives of the project have been attained.

*Director, Projects Department (Latin America and the Caribbean), World Bank, Washington

3. This Paper deals with some aspects of the activities that take place before the start of construction, i.e. what the Bank does and what others do in the process of initiating and preparing a large capital project and in securing the financial resources for carrying it out.

Choosing a project

4. The Bank lends the bulk of its funds for specific projects. It wants to assure itself that the projects it finances are economically and technically sound, but its objectives are much broader. As a development institution it is concerned with assisting its developing member countries in achieving sustained economic growth with a wide distribution of benefits. To achieve this goal, the Bank engages in a dialogue with its borrowing countries on a wide range of economic and development issues, including the country's public sector investment plan. Only when it has formed a view of the country's investment priorities does it select, in consultation with the government, specific projects in various sectors for which it will consider making loans.

5. The process of project selection usually takes place in the framework of sector analysis which seeks to identify an optimum strategy for the development of a sector, such as transportation, energy or agriculture, and to determine the investment priorities for particular projects in that sector. Such sector work is carried out by Bank staff, often with the support of individual consultants, by specialized UN agencies such as FAO, WHO, UNESCO, etc., with whom the Bank has co-operative agreements, or by private consulting firms financed by the Bank or by the United Nations Development Program (UNDP), with the Bank acting as executing agency.

6. The specific form in which the Bank and the government of the country concerned are involved in this sector work varies with the contractual arrangements and the source of funds. Generally the Bank has an important input in the design of the study, its conduct and staffing and the evaluation of its results. When it acts as executing agency for the UNDP the Bank itself selects and negotiates a contract with a private firm or team of consultants. In other cases this is the responsibility of the government agency concerned, in consultation with the Bank.

7. When a high priority project has been identified through this sector work and the Bank has agreed to consider lending for it, the task of project preparation begins. In the case of large capital projects this generally means the commissioning of an engineering and economic feasibility study.

Project preparation

8. Even though a project has been identified as being of high priority its size, location, cost and economic and financial feasibility may not yet be known. The sector work may have identified the urgent need for additional electric power or for doubling the capacity of a port, but may have given only a general indication of how and at what cost this could be accomplished. Considerable engineering work is usually required to decide on the optimum location and technical characteristics of the project. In some cases a government agency or semi-autonomous entity is in a position to undertake this feasibility study work with its own staff, but in most developing countries the help of consulting firms, whether domestic or foreign, will be needed.

9. The Bank's role at this point is in the first instance to agree the scope of the work with the prospective borrower. This should result in 'terms of reference' which spell out the understanding between the Bank and the borrower of the objectives and scope of the required services, serve as the basis for inviting proposals from consulting firms and may later become part of the contract with the selected consultants. As in the case of sector work, the Bank may finance such feasibility studies through a loan or credit or, as executing agency, administer study funds made available by UNDP.

10. In their choice of consultants the Bank and the borrower are guided by two considerations. The first is to get the work done expeditiously and to high standards of quality. The second is to take advantage of the opportunity to strengthen local capabilities so that in the future the country itself can undertake a larger share of the feasibility study. These objectives may, at least in the short run, be partially conflicting. If there are qualified local consultants capable of doing all or part of the work, the Bank encourages their use as part of its general interest in developing local capabilities. If, as is often the case, the necessary expertise is not available

locally and foreign consultants are needed, the Bank is interested in exploring opportunities for on-the-job training of borrower's staff through appropriate provisions in the consultants' terms of reference. On the other hand, the Bank does not endorse a strategy of developing local capabilities that would be at the expense of the quality of the end product.

11. The actual selection and hiring of the consultants is normally done by the borrower subject to the Bank's concurrence on the terms of reference, the qualifications of the consultants for the task at hand and the contractual arrangements between the borrower and the consultants. Throughout the course of the feasibility study the Bank remains in touch with the borrower and the consultants to follow the progress of the work and help resolve any technical or other issues that may arise.

Estimating project costs

12. Normally a feasibility study will advance the definition of a project to a point where decisions on the commitment of additional financial resources can be made. In large capital projects the question often arises as to whether the cost estimates made as part of a feasibility study are sufficiently reliable to form the basis for drawing up a financing plan and concluding a loan agreement for the project, or whether additional engineering and design work will be necessary. Bank experience has shown that in the case of projects, such as highways and irrigation canals, that may cross many types of terrain over large distances, more detailed engineering is normally needed to end up with reliable cost estimates.

13. For large dams and ports where construction is confined to a relatively small area that can be more easily investigated, reliable cost estimates may be obtained on the basis of the preliminary engineering work of a feasibility study. However, caution is required when geological investigations and sub-soil explorations indicate that conditions cannot be expected to be uniform over the construction site. In such cases additional exploratory and design work will tend to pay off in reducing the possibility of unpleasant surprises. Large cost overruns in dam or port structure foundation work, rock excavation for tunnels, or dredging of port basins can often be traced back to inadequate sub-soil exploration

work. Cost estimates for thermal power stations and industrial plants generally are not as vulnerable to the hazards of inadequate sub-soil information, as the cost of their foundations is typically only a small percentage of their total cost.

14. Other uncertainties with respect to the cost of large capital projects may arise in certain developing countries, because projects of a similar nature and magnitude have not previously been constructed there and economic or political instability make it risky for contractors to enter into long term commitments. Contractors may be reluctant to bid on projects in such countries and, if they do, they will want to cover their risks as they see them. This makes the outcome of the bidding process very unpredictable. In such circumstances it may be advisable to wait for the bidding results before entering into commitments for the financing of a project. Generally, however, the cost estimate made by engineering consultants forms the basis for the substantial financial commitments necessary for constructing large capital projects.

15. Before the Bank enters into loan negotiations, it undertakes an appraisal — a final, formal analysis of the project in which it considers its economic, technical, managerial, organizational, commercial and financial aspects. A review of the engineer's cost estimates is a vital part of that appraisal. It is normally done by Bank staff but for some very large projects the services of specialized cost estimating firms are sometimes used as well.

16. In the first instance this review focuses on the engineer's base cost estimate — the cost of the project at today's prices. In an inflationary environment, however, the estimation of future price increases is of crucial importance in arriving at a reliable cost estimate, particularly for large projects which take many years to complete. Even when prices are expected to rise at modest annual rates, the effect of rising prices on the total cost of a project during a long construction period can be quite severe. When high rates of inflation are predicted, the time element becomes a major factor in the calculation. A realistic assessment of the total time to completion as well as the construction progress profile then becomes essential to reliable cost estimates. Bank experience indicates that many major cost overruns tend to be directly related to over-optimistic predictions of the pace of construction progress. In developing countries allowances often have to be made for circumstances that will make it difficult to attain the same standards

of efficiency that are normally achieved in industrialized countries.

17. In estimating expected future price increases, a distinction may have to be made between domestic and world-wide inflationary trends. Whether or not domestic or foreign prices for particular types of work or equipment will be in line with general trends should also be evaluated. For example, if the construction industry in a certain field is currently over-extended or depressed, this should be taken into account in estimating bid prices. Large projects sometimes create their own inflation; local resources such as land, labour and raw materials may increase in cost as a direct result of, or in anticipation of, the project.

18. It is important for both the borrower and the Bank to know the project's foreign exchange costs. In reviewing cost estimates, the Bank pays special attention to this aspect to make sure that the foreign exchange cost estimate not only includes the direct use of foreign resources for the procurement of foreign goods and services, but also their indirect use by foreign and local contractors, suppliers and consultants.

19. In most cases cost estimates for Bank-financed projects are based on the premise that international competitive bidding will be used. This is not only usually the best method to ensure economy and efficiency in the execution of the project, but it also gives all the Bank's member countries, developed and developing, a chance to compete in providing equipment, materials or construction services for the works it finances. In addition, the Bank, as a development institution, is interested in encouraging the development of local contractors and manufacturers. It therefore grants, under certain conditions, a limited margin of preference to domestic manufacturers and, in some low income countries, to civil works contractors, when comparing domestic with foreign bids.

Financing the project

20. The estimate of the total cost of the project and the expected annual need for foreign and local currencies during the construction period forms the basis for drawing up a financing plan and deciding on the size of the Bank loan. Although there are circumstances under which the Bank agrees to finance part of the local costs of a project, it has a strong preference for limiting its contribution to

no more than the project's foreign exchange requirements.

21. The balance of the funds will have to come from other sources and it is necessary to insure that such funds will be available on reasonable terms. In some cases such as a loan to a Government for a highway or an irrigation project, this may involve no more than an undertaking by the Government to provide the funds from budgetary allocations. For loans to revenue earning entities such as industrial plants, power companies, railways or autonomous port authorities, most or all of the financial requirements, including the servicing of the Bank loan, normally have to be covered by the enterprise. These requirements may include, in addition to the project costs, interest during construction, repayment of instalments of existing debts and other capital expenditures.

22. Possible sources of funds to complement the Bank loan are retained profits and depreciation funds, loans from suppliers or other lenders, the proceeds of equity issues or capital contributions by the Government. Detailed financial forecasts including income and expenditure accounts, *pro forma* balance sheets and cash flow statements will have to be prepared, covering as a minimum the project construction period, to make sure that the funds will be available when needed.

23. The foreign exchange requirements of large capital projects are often so great that the funds the Bank can make available are not sufficient, and other foreign lenders have to be approached. The Bank's involvement in helping the borrower make arrangements with other lenders varies a great deal from case to case depending on the wishes of the borrower and the co-lender, and their previous experience. In the case of co-financing with export credit agencies, the borrower typically takes the responsibility for seeking the best possible price from suppliers, combined with the most advantageous terms from the export credit agency. The Bank's role is usually limited to splitting the project in such a way that, for example, civil works contracts can be financed by the Bank and equipment purchased through export credits.

24. The largest sources of co-financing are the official sources of aid, bilateral programmes as well as multilateral institutions. In many cases the borrower will take the initiative in suggesting possible co-lenders from this group; in others it will ask the Bank to help identify possible co-lenders and to brief them on the project. Once co-lenders have been agreed on by the borrower, the

Bank works closely with them. Some may wish to participate in the Bank's appraisal, others may prefer to carry out their own independent project analysis. In some large projects many co-lenders may be involved and the Bank often acts as the co-ordinator and spokesman of the group.

25. In the last few years, the Bank has been actively seeking participation by private banks in co-financing arrangements for some of its projects. Private bank funds are available mostly at medium-term (5—8 years), which may not be suitable for the financing plan of large capital projects. The Bank may, however, be willing to lessen the burden on the borrower by foregoing short maturities on its loan and skewing its own amortization schedule toward the later maturities to accommodate private loans, if their terms are sufficiently attractive.

26. This brief Paper could only touch upon a few of the most important issues in which the Bank becomes involved in the initial stages of large capital projects. The issues are many and complex and to deal with them requires people of many professional backgrounds such as economists, financial analysts, lawyers and many kinds of technical specialists. Among the technical specialists it is often the civil engineer who plays a key role long before the start of construction.

Discussion on Papers 1-4

Mr Taylor

1. The purpose of Paper 2 is to serve as an *aide-mémoire* to the full range of facilities ECGD can offer contractors engaged in overseas projects. Contractors should consult the Department at the earliest possible stage when considering bidding for a project. Although offering a wide range of facilities there are, nevertheless, occasions when the Department will not be able to help: cover on some overseas markets may have to be restricted because the risk of non-payment is judged to be unacceptably high (although the UK's national interest, often seen in terms of providing worthwhile employment opportunities, may be taken into account before an application is declined); and ECGD cannot match 'aid' terms that may be on offer from the competition (although it may be possible to consider matching mixed credits where the commercial terms element predominates).

2. The scope of the new cover to be provided against 'joint and several' risks (paragraph 19) available selectively for projects valued at £50 million or more, goes beyond its predecessor — the project participants insolvency cover facility. It provides cover for 80% of losses incurred from factors outside a contractor's control, up to a maximum of 20% of the UK value of the project, which arise from the shortcomings of UK partners or subcontractors and are not recoverable contractually but which do not necessarily have to be associated with the latter's insolvency. This facility should help UK contractors to offer more competitive bids for large overseas projects.

Mr Owen

3. Traditionally, buyers have tended to see the major factors in awarding a contract as price, technology, delivery, etc. The currency of the contract and the loan have been a relatively low item on the list. Yet if one looks over the last four or five years at the contracts which have been placed in Deutschmarks, sterling, dollars, and so on, and looks now at what it has actually cost the people who have placed those contracts to do so in those relevant countries and currencies, the difference which emerges in the cost of the particular project or equipment has been as much as 80%.

4. Fortunately, the UK now has an export credit system which allows greater flexibility in the currency of ECGD-backed loans than the equivalent arrangements in almost any other country in the world. This is a great advantage to UK contractors and a sales tool which can be used effectively, even though it involves a certain amount of complication and banking technology.

5. The way in which this tool is best used varies considerably from country to country and buyer to buyer, and it is not easy to generalize. Some buyers still want to borrow and contract in sterling, at all costs. Generally, with some limitations on size, this is now feasible. Many buyers, however, actively prefer to contract with the UK in US dollars, often because their own currencies are linked to the US dollar. There are even a few countries where it may be possible to make an ECGD loan available in the local currency itself – Hong Kong, Singapore, Malaysia, and a few others – and this may be a plus factor for the British bidder compared with his competitors who do not have this ability, in a situation where the borrower is keen to match the currency of his borrowing with that of his revenue.

6. In other situations, the ability of the exporter to contract in a strong currency, for example, Deutschmarks or Swiss francs, and sell his receivables forward in the foreign exchange markets over the project construction period, to produce a significant price discount, can be an important factor in winning a contract. This has happened on one or two occasions in the past. It is particularly relevant where there is competition from a contractor bidding a lower price in a hard currency out of a country with a low rate of inflation. It is relevant also where, as still happens with remarkable frequency, the buyer is insensitive to the currency factors and

simply compares the prices of different bidders at the spot rate of exchange ruling on the day of the tender, without taking into account the forward weakness or strength of the respective currencies. In such cases, the price quoted by, say, a German or Japanese company in Deutschmarks or yen may, for example, appear five or ten per cent cheaper than the UK price. In this situation the UK contractor does, with the new ECGD facilities, have the option of converting his bid into Deutschmarks, using the forward foreign exchange market and thereby reducing his price by a factor which will depend on the length of construction period but, in the case of the Deutschmark, should today be anything from seven to twelve per cent on the contract price. Then the buyer still has a situation where he is comparing bids on an equal basis, currency against currency, and the UK is cheaper.

7. In situations where a British contractor does bid in a currency other than sterling there are a number of contractual risks to which he is potentially exposed and which his banker must seek to minimize, with the help of ECGD. The first risk is that, having dealt forward on the foreign exchange market, his commercial contract may fail to become effective or, having become effective, it may be terminated by the buyer. Both risks are covered by the ECGD insurance which goes with a currency financed contract at no extra cost in premium.

8. The second risk is that of late delivery. Having taken out forward cover to a particular date, the exporter finds that he does not deliver on that day but three or four months later, in which case he will have to roll over his forward foreign exchange contract. This particular risk can be covered by the use of the forward foreign exchange market on the basis of an option contract (which gives the exporter a period of usually up to six months during which he may deliver the relevant foreign currency).

9. The third risk, which does not apply in a fixed price situation but does apply in a contract with an escalation provision, is that the contractor does not know in advance the exact amount of escalation payments he will be claiming and can therefore deal forward only for an approximate amount. If he is wrong in his estimate he will carry the risk on the margin of his error. There is no simple solution to this problem yet, although there are ways in which it can be minimized to the point where it does not cause concern to the majority of contractors.

10. Finally, and most important, there is the risk of an adverse movement in the forward foreign exchange rates between the date of bidding and the date of award of the contract. The exporter will have made his bid on a particular day, based on the forward foreign exchange rates applicable at that time. There is no guarantee that the forward rates on the day he signs the contract will be the same. Indeed, it is highly likely that they will be different. To deal with this situation the ECGD have introduced an insurance scheme designed to cover such a risk — the so-called 'tender to contract' insurance cover. Under this scheme the contractor bases his bid on the forward rates applying on the day he puts in his tender. ECGD guarantees him against an adverse movement of anything more than three per cent in the forward rates between the date of tender and the day on which he actually deals forward in the foreign exchange market: that is to say, after he has signed his contract. If the forward rates move in the exporter's favour — in other words, if sterling becomes weaker — ECGD take the benefit of any movement of more than three per cent. There is an insurance premium charged by ECGD for this which is not insubstantial but is nevertheless not so serious as to destroy the potential advantages of foreign currency financing.

Mr D.J. Westall (*Director, CJB Earl & Wright Ltd*)

11. A project runs the risk of one partner or supplier of critical items failing to perform. In a large project there is usually a critical item: for example, mainline pumps on pipelines, winding gear in mines and container cranes in ports. In the event of that critical item or group of items being delayed, or failing to meet its performance guarantee, there is the danger that the penalties applicable to the whole project, either on time or performance, will be invoked by the client.

12. Another example of this problem is in a petrochemical scheme: if the cracker does not crack then the downstream plants, which are probably provided by other partners, cannot perform.

13. The company supplying these critical items, without which the scheme will not work, is often relatively small and may be of insufficient wealth to indemnify the joint venture against failure.

It may also find it impossible to obtain any insurance to help. There seems to be no obvious way in which this situation is likely to alter or be remedied.

14. The new ECGD facility touches on a related but different point: namely, the circumstance of a partner failing financially. This is a different kind of failure, and the cover mentioned in Paper 2 was limited to 20%. A greater problem is that of a partner or supplier getting into, for example, industrial relations difficulties involving substantial delay, or having a catastrophic fire in his factory.

15. How can one cope with the failure of a relatively minor partner, supplying a critical sector of the project, creating the risk of invoking the full guarantee penalties against the whole project?

Mr R.L. Fitt *(Sir Alexander Gibb & Partners)*

16. Is there any move towards dealing with the situation where you have a consortium made up of contracting bodies from different countries?

Mr E.M. Gosschalk *(Associate, Sir William Halcrow & Partners)*

17. The following remarks point to the difficulties of making decisions to implement projects.

18. Fortunately we are not faced everywhere with the situation which exists in Indonesia, where it is not permitted to place a local contract which is intended to exceed a period of one year. The amount of finance to be available for that year is decided only three months before the work is due to begin. Consequently, there is very little assurance on continuity of work or workmen, or of the date when the project is likely to be completed.

19. I worked on the planning of certain hydroelectric projects in Northern Ireland. These projects were first thought of in about 1922 and were pursued for more than thirty years. They were scrapped in about 1954, when a nuclear power alternative seemed imminently available. Had these projects been built at any time since 1922, Northern Ireland as a whole would have been more than delighted with them today.

20. Individuals in democratic countries have lost the nerve to make decisions. More and more they seem to feel the need to fall back on committees, economists and computers, to make decisions for them. In fact all these fail because it is not known on what criteria to base decisions. The real need is for crystal balls.

21. Economists are, I believe, in disarray. It is often not so much a question of 'economist' but more of 'e-con-em-most'. At the recent Colston symposium, the principal economist speaker gave a paper[1] on the evaluation of tidal power projects which was a brave exposure of the uncertainty and doubt in economics. He remarked on the impossibility of deciding what a test discount rate should be between 10% and zero, and on the impossibility of having to weight the benefits in the cost benefit equation for the poor as compared with those more affluent. The test discount rate requires an impossible judgment on the importance of the future compared with the present day, while the weighting of benefits requires an impossible judgment on how much of those benefits are likely to be passed on by the richer to the poor. Therefore it is not surprising that when faced with the question of how a rational decision is to be made whether to proceed with the Severn Barrage, the answer of the economist was that he did not know.

22. If economics cannot produce decisions for us, can we turn perhaps to sociology? Most of us would prefer to face the risks and uncertainties of economics rather than place ourselves in the hands of sociologists. Nevertheless, the time has come when human instinct, which has fallen into disrepute as a decision-making medium, is due for a revival of respect. After all, human instincts to progress and make heroic decisions, sometimes to succeed and sometimes to fail, are part of our nature. There is an important element of subjective judgment in all decision making, fallible as it may be.

23. There is an attraction in energy accounting methods, but they could lead to counting neutrons and to the impossible problem of weighting the value of energy that is realizable now compared with that which might be realizable in the future; also to facing the problem of assessing the energy value of the modern equivalent of a Rembrandt.

24. The future of decision making depends on more attention being paid to civil engineers. There are three basic questions on

which we should concentrate: Is the project technically sound? How much is it wanted? Can the money for it be found?

Mr A.R. Parish *(Deputy Chairman, W.S. Atkins Group Consultants)*

25. I have had the impression that the World Bank is backing away from large projects. Is this in fact true?

26. Mr Owen recommends the negotiated package, and Mr van der Meer enunciates the Bank's policy and generally commends international competitive bidding. Can the Authors produce any facts in support of their opposed contentions?

27. As to the World Bank's position, one understands that this is enshrined in some of the documentation of the Bank. However, I wonder if this owes more to the interests of the supplying countries than to the interests of the project and the buyers. After all, the Bank is faced with a group of executive directors in Washington representing, at least in part, the countries which may wish to gain the projects or part of them. Is this devotion to international competitive bidding not due to the fact that it gives the Bank's civil service a ready answer to the executive director, who is complaining that his country has lost a contract?

28. Contracting in weak currencies obviously gives a project owner an advantage, as long as it remains weak over the whole repayment period of the loan involved. In my experience, the World Bank will not allow a project appraisal other than on the basis of the exchange rates that exist at the instant that the decision on that appraisal is to be taken. One is not allowed to take any notice of an opinion of the future strength or weakness of the various currencies, even when a forward market in those currencies exists.

29. I have been given two explanations of the reasons why the Bank takes this view. The first one is that the forward markets are not a true reflection of the likely commercial rates of exchange in the future, but more a reflection of differential interest rates between countries and, possibly, more a playground for speculators than a true commercial operation. The second view is that it would be rather embarrassing for the Bank if it approved appraisals which contained views of the likely future rate of exchange of members' currencies, because in a sense that is approving a view of the possible

future strength or weakness of the economy of a member country of the Bank. Have these points any truth in them?

30. As far as I am aware, a borrower from the World Bank has no choice in the currencies that he is lent. The Bank will disburse in whatever currency the contract may require, but in that case unless a loan is being made in the currency that is wanted, the borrower has no choice in the currencies that are being lent to him. All the Bank is doing is acting on behalf of the borrower to exchange one currency for another on the commercial exchange markets. The borrower is likely to get strong currencies, because it is in the strong currency markets that the Bank is likely to be able to sell its bonds.

31. Does the British Government stand as firmly behind its exporters as other countries' governments stand behind their exporters? In many countries there is not the complete distinction made between trade and aid that is the British Government's practice. They are muddled up and the *crédit mixte* is a classic example of this — to the advantage, in my view, of the foreign exporter. In Britain we seem to keep them separately compartmented, to the point of regarding any aid which helps British trade as being somewhat indecent. How are the British going to match the *crédit mixte*?

Mr J.T. Edwards *(Partner, Freeman Fox & Partners)*

32. Apart from the problems of a British contractor bidding in a currency other than sterling there is also the problem of the client and the engineer who have to adjudicate on such tenders. In large projects which are not financed by the World Bank or other similar bodies, it is common practice for the tenderer to be required to offer finance for part of or the whole of his offer, when tenders are invited from contractors. In a number of countries this leads to offers of finance which may be partly in the country in which the project has to be executed and partly in the tenderer's own country, or in the currency of a third country. If the tenderer is a multinational consortium, the tenderer's portion may be in two or three currencies.

33. In assessing the tenders, it is therefore necessary to take into account, in addition to technical factors and a direct financial

assessment, the interest rate, length of credit, starting date of repayment, currency depreciation or appreciation, and rate of inflation.

34. An example of this type of project is the Modified Initial System of the Hong Kong Metro, for which Freeman Fox & Partners are the consulting engineers to the Hong Kong Mass Transit Railway Corporation. In order that tenders could be fully compared, it was necessary to develop a tender assessment program, so that tenders could be brought to a common base, using a computer for the calculation. This tender assessment program (TAP)

(a) prepares a cash flow from the tender program and cost data
(b) splits the cash flow in the proportion of the tender currencies
(c) escalates costs at assumed rates
(d) incorporates assumed currency exchange rate fluctuations
(e) determines the loan draw down on several bases
(f) allows for capital repayment and capitalization of interest on several bases
(g) includes fixed and variable financing fees
(h) determines the net present values at a number of discount rates.

By this procedure an overall view of several tenders can be obtained, making a wide range of assumptions of financial trends. Thus an opinion can be formed of the relative merits of tenders, taking both the tender price and the financial offer into account. This development of complex financial offers as part of the tender appears to be a growing feature of the large capital project. What are the Authors' views on this?

Mr Taylor

35. The question in paragraph 15 has been exercising both Government and industry for quite some time and this is evidenced by the number of study groups which have examined various aspects of it in the last few years. In the smaller projects contractors can generally live with the problem. It is in the larger ones that the risk

may become insupportable, where penalties may well exceed the net worth of the participants. We therefore hope that new development of ECGD's Project Participants Insolvency Cover — the 'Joint and Several' Indemnity — will match the situation.

36. Essentially this takes over at the point at which the sub-contractor's contractual indemnity to the main contractor leaves off. There will normally be some provision in a contract for indemnification by the party which fails to perform, but this is almost certain to be limited in amount. No sub-contractor could provide total indemnification if the whole project fails because of his own default.

37. Let us assume that the sub-contractor can, and has to, produce 10% of his share of it by way of indemnification, but because of the domino effect 10% is not enough to cover the overall liability. Though the Joint and Several Indemnity, as Mr Westall said, is limited to a total maximum liability of 20%, this is 20% of the value of the entire UK content of the contract and within this ECGD would bear up to 80% of the ascertainable overrun which the UK partners have incurred.

38. With reference to paragraph 16, given that the Joint and Several facility is intended for large multi-discipline projects and is to be given on a selective basis for contracts which are judged to be high in the national interest, the question raises some difficulties.

39 How far do we go in appearing to encourage a foreign sub-contractor, while some British sub-contractors may want and have need of the job? How far could ECGD, a UK Government Department, pursue a recourse claim against a defaulting, but not as yet insolvent, overseas company? The option of having insurance cover against the consequences of insolvency of a foreign partner or sub-contractor remains open under the Project Participants Insolvency Cover. But for the present, the wider Joint and Several cover, going beyond the insolvency situation, is being restricted to UK parties — UK partners and UK sub-contractors.

40. The scheme could also be made available where a large UK firm, acting as a major sub-contractor to a foreign project leader and involved on the scale for which the facility is intended, wishes to have cover against his own UK sub-contractors.

41. Mr Parish questioned the attitude of the UK Government to *crédit mixte*. We have had the powers for some time to deploy mixed credit facilities, although in practice they have not yet been used

and we have tried to maintain a clear distinction between commercial credit and aid. Nevertheless, we are concerned at the way competitor countries offer *crédit mixte* and similar facilities, and have therefore introduced more flexibility into the administration of our bilateral aid programme to give higher priority to the commercial importance of a number of developmentally sound projects identified in developing countries for which there is no aid allocation available or where it is already fully committed. Thus mixed credits might be offered to make sure, in selective cases, that UK exporters are not left behind by their competitors.

42. It is internationally agreed that once the aid element of a particular case produces a degree of 'softness' which involves a grant element of at least 25% then the case should be regarded more as forming part of an aid programme and not, therefore, susceptible to matching by officially supported commercial credit.

43. We are now much better informed about other countries' practices in this field and are therefore in a better position to match their support where appropriate or indeed to take the initiative in using the *crédit mixte* approach.

Mr Owen

44. Mr Parish has, I think, in his summary sharpened up the position of Mr van der Meer and myself. Whether it is better to have a negotiated bid or international competitive tender depends on many factors, including the nature of the project. For large projects, particularly in the industrial sphere, where the time factor and the cost inflation factor are great, there is a strong argument for the negotiated route. As examples of the different approaches I would mention the Sicartsa steel plant in Mexico (competitive tender) and compare it with the more successful handling of the Altos Hornos de Mexico project; a similar contrast exists between the handling of CSN's expansion programme in Brazil and the Acominas plant.

45. With regard to currency factors, I was interested in Mr Edwards' comments and delighted to hear about his evaluation system. I hope this is used in as many cases as possible. At present such sophisticated means of evaluating tenders seem to be used in a small minority of cases. The vast majority, in my experience,

still involve a straightforward comparison of prices, usually on a spot basis with the currency factor treated as marginal, if it is taken into account at all. Nevertheless even with a computer program there are still a number of elements which represent judgement rather than objective data, the most important of which is the relative exchange rate movement over the period not just of the commercial contract, but also of the loan which may be as long as 10–15 years.

46. Dealing with Mr Parish's question on what the forward foreign exchange market reflects, the answer is that as far as it goes – up to about 5 years in dollars versus sterling and a shorter period in sterling versus most other currencies – it reflects primarily the differential in interest rates between currencies. These interest rates, however, are of course influenced by peoples opinions about the strength and weakness of different currencies and, above all, of relative rates of inflation in the countries concerned.

47. I assume that Mr Edwards, when he does his analysis, works on the basis of the estimated future inflation rates in different countries, applies that to the currency of each tender, and comes out with a factor which is applied to the contract price reflecting the expected forward exchange rate movement. To the extent that this is done it will tend to be a major advantage for UK bidders who, generally speaking, are bidding in a weaker currency than most of their competitors. Thus, if these currency factors are fed in, it will tend to assist the UK contractor.

48. From the point of view of the UK contractor, there is an element of certainty in the UK currency export credit scheme. Whatever opinions may exist about future exchange rates, it is an objective fact that the UK contractor can sell his foreign exchange receivables forward and realize a specific and known amount of sterling which he can translate into a specific contract price. This is easier for the UK contractor than for the foreign borrower or buyer in as much as the contractor has access to the Bank of England as a 'counter party of last resort'. In other words, if the commercial market in the relevant currency does not extend far enough into the future or is too thin, the Bank of England will step in to create a market. This means that the UK contractor is in a position to take on the responsibility of dealing forward in a situation where the foreign buyer may not have that capability.

49. Looking at the period beyond which a commercial foreign exchange market exists, I agree with Mr Edwards that one should not say that because it is not known what exchange rates are going to be one should not try to take them into account. Currency obviously remains a very important factor and as sophisticated an attempt as possible should be made to estimate its effect.

Mr Yassukovich

50. The increasing practice of asking contractors or operators of large projects to tender with an attached financial package — a point raised by Mr Edwards — is largely misunderstood, or is misinterpreted perhaps by those who receive, arbitrate or interpret the competitive tenders with financial packages attached.

51. The European Banking Company and many other banks are constantly being approached by contractors or operators seeking a firmly committed financial package to accompany a tender. Most of the time there is insufficient information available to make or to organize such a firm financial commitment. A truly firm commitment as opposed to an indication or expression of opinion as to what the financing costs will be, from a bank or an insurance company, has to be remunerated with a commitment commission or a premium. Bidding tenderers are not normally in a position to pay that remuneration for the period required while they wait to see whether they are successful or not.

52. Invariably the financial packages which accompany tenders of this type are dressed up as being far more firm than they actually are. They are usually of such a conditional nature that any change in circumstances goes beyond the change in circumstances of the project itself. But changes in economic or market conditions usually allow the lenders, or other providers of finance, to alter the terms and conditions.

53. Highly educated guesses can be made about the terms and conditions which will apply to the equipment-related finance, because the current prices quoted by the agencies are known, but the agencies are not willing to make firm commitments on a theoretical basis. It is very much a question of communications rather than firm offers. This is not always appreciated by those receiving these offers.

54. The value of the financial package as part of the tender is also exaggerated. A lot of people tendering in a competitive situation are surprised to find that the financial communications they receive are almost identical. This is simply because the finance is coming from the same market and therefore different banks will assess the conditions of that market in exactly the same way. The trend of asking contractors or operators of projects to attach firm and totally wrapped-up financial packages does not seem useful, because it does not take into account the fact that market conditions are too volatile to allow those packages to be firmly committed.

Mr van der Meer

55. Mr Gosschalk mentioned that decisions about energy projects cannot be made. When we in the Bank consider energy projects we try to make an assessment of the economic value of these projects, but basically we are unable to determine what the full benefits of an energy project are. We can and do look at the willingness and ability of the beneficiaries to pay. We use that as a proxy for economic benefits, which we think are usually far in excess of that willingness to pay. However we often get into prolonged discussions with our borrowers in trying to have them make the decision to impose adequate charges. Tariff increases in electric power tend to become sensitive political issues.

56. The problems with which we are faced are usually of that nature, and not related to our ability to decide with which project we should go ahead. Given the need for energy, we seek the least-cost alternative of providing it, even though we do not know what the value of energy will be in the future. This future value will be the same for any alternative means of providing energy, so it is not too difficult from that point of view to make a choice between a tidal plant, a hydroelectric plant or a nuclear or fossil fuel plant. In choosing between alternatives, we have to make predictions about future construction costs and fuel prices. But the difficulty of making these assessments is somewhat exaggerated.

57. The main problem we face in getting such capital intensive projects started is the provision of funds. Some of these projects are so large that bringing money together to undertake them

often becomes the bottleneck. An example is a project we are considering in South America which will cost some $4000 million to $5000 million. The benefits of that project are not in question, but the question is: where does one get the resources? The Bank will be able to finance only a small part of the cost and we are actively working with Government authorities to try to induce others to participate financially in this venture.

58. Recently there has been quite a change in emphasis in the criteria that the Bank uses in deciding on the kinds of projects it would invest in. This is very much an emphasis in favour of directly trying to help the poor in our member countries, rather than relying solely on what is known as the 'trickle-down' theory of economic development. The Bank has done a lot in the spheres of agricultural development and what we call rural development projects, and has, in other sectors, shifted its emphasis away from simply providing the basic infrastructure to looking for ways to reach the poor and economically powerless.

59. For example, in roads we no longer just finance the large highway from A to B, or parts of the primary national network, but increasingly finance feeder roads which may be far away from the main arteries of traffic but serve and stimulate production by the small farmer.

60. In energy we tend now to look at distribution, rather than just generation and transmission, and particularly at the circumstances under which distribution takes place. Does it reach only the well-to-do customers? Are there ways in which we can make sure distribution also reaches those who may not be the most favoured clients of the power company? Is there a way of ensuring that they can afford that service? Do we need cross-subsidization to meet the basic energy requirements of the poor?

61. The same is true in water supply. We do not just finance the development of the water source, or the filtration plant or the large transmission lines. But we look at distribution and consider where the network goes. Does it reach the poor areas of the city? What can we do about water supply in isolated rural areas?

62. Having said that, it is still true that about half of the money that the Bank lent last year, for example, which was about $7000 million, went for what we have called in this conference 'large capital projects' which are largely in the 'traditional' sectors. In fact we are making new overtures in these sectors. We have long

been lending for power and transportation but we have not lent much in the past in the fossil fuel field, particularly not for oil development. We have financed power plants that were oil or gas fired and financed oil and gas pipelines, but never anything in the exploration or development of oilfields — largely because other finance was usually available.

63. We are now committed to a policy where we are actively seeking opportunities to invest in the development of oil resources in less developed countries. We have made one such loan in India, through our participation in the Bombay High oil and gas development. There are a number of other oil and gas projects being considered. Last week we made our first loan for the petrochemical industry in Brazil, to a company called Copesul, where we contributed $85 million to a project with a total cost of around $500 million. It is the first part of a petrochemical complex, also requiring 'downstream' units, which will probably need investments of a similar amount, before the whole complex is operational. This is something that we have not done before.

64. Although it is true that we pay much more attention to who the beneficiaries of bank loans are and how we can induce Governments to reach the disadvantaged parts of society, we do not neglect the large projects needed for the provision of energy, water, transportation and basic industry.

65. There is a difference between engaging consulting services and services of contractors in our project work. So far as consultancy is concerned, to the extent that we are directly involved, which is in a small percentage of cases, we try to achieve a reasonable spread of the work among countries. These are the cases in which we act as executing agency for the United Nations Development Programme.

66. When it comes to construction we have a policy of requiring international competitive bidding, where this is feasible. There, of course, we do not particularly look for a spread of work among countries, but for the lowest evaluated bid. The awards might all fall in the UK, and this would not worry us. Awards go to whoever can supply the service at the lowest cost. Whether this is better for the borrower than a negotiated contract is not always clear. There have been a number of cases however, where tied procurement, for instance, has led to much higher costs than the system of international competitive bidding. Some of these cases are well

documented by the fact that when the borrower, faced with the high cost of tied procurement, decided not to take that option and went for international competitive bidding, he found that he profited handsomely from the change.

67. There are cases (particularly in large industrial projects) where the turnkey contract would be advantageous, particularly because of the time factor, over trying to make arrangements for separate elements through international competitive bidding. The Bank is fairly flexible in that respect. In cases where it seems that the co-ordination of certain contractual activities is crucial to the success of the project we will agree to the turnkey approach under a single contract, awarded after international competition.

68. Regarding the question of what the Bank's policy is on international competitive bidding and why we do not take into account the type of currency in which the bid is expressed, two explanations have already been offered: that it would be unseemly for us to predict the future value of our members' currencies, and that the forward markets might not be a true reflection of the strength or weakness of the currency.

69. In this question of award of contract under Bank loans, we retain the right to object to an award of contract that in our view does not adhere to the principles of international competitive bidding. These objections can become a hot issue with the representatives of the member country which is on the losing end. We obviously have to be as fair and objective as we possibly can be in making these assessments. This would be very complicated if we took into account what I would consider to be of necessity a rather subjective view of what a particular currency would be worth in the future. The practice of the Bank is to use the exchange rate of the date the award decision is made.

Reference

1 LECOMBER J.R.C. *The evaluation of tidal power projects.* Colston Research Society 1978 Symposium. University of Bristol, 1978

5. Influence of the nature, size and location of projects on organizations

A.G. Frame, BSc, MA, FIMechE [*]

Introduction

1. Irrespective of the nature, size and location of the capital project, strong client guidance and control is essential. This Paper discusses the roles of client and engineer in the development of capital projects of varying sizes in different locations. In many capital projects there is no alternative choice of location — obviously mines have to be developed where ore-bodies are situated. It is perhaps unfortunate that many of the world's great ore-bodies are located in remote areas, more often than not in undeveloped countries. Participation in such areas increases the already significant problems of organizing a large mining project.

2. The key factors in the development of a project in a remote location are discussed in the following paragraphs.

Project organization

3. The client must set up an organization to develop the project at an early stage; furthermore, the head of this team must be responsible for representing the client with all other parties, and must be vested with clearly defined delegated authorities to enable him to manage the project in an efficient manner. Although the client will almost certainly have a Board of Directors or Management Committee responsible for overall policy on the project the Board should not interfere, and it must ensure that the head (or Project Manager) of the client's project organization has these clearly defined terms of reference and delegated powers. In addition, the head of the client's organization should be capable of

---

[*]Deputy Chief Executive, Rio Tinto-Zinc Corporation

holding a senior position, preferably the top one, in the completed operation.

4. The client's project team must have within it people capable of developing mine planning, optimizing the rate of mining, deciding on the method of beneficiation, etc. This latter process will inevitably involve the design and construction of a pilot plant to treat a bulk sample of correctly prepared run-of-mine ore.

5. Since it is extremely unlikely that the client will have the full range of professional disciplines necessary to develop the scheme within his own organization, the selection of the engineering consultant (engineer) to develop the feasibility of the project is of prime importance, and invariably the same Consultant will develop the total project under the aegis of the Client. The general practice in the mining and metallurgical industry is for the client's group referred to above to prepare the scope of work to be performed by the engineer, and then request proposals on, *inter alia,* the names and availability of key staff whom the engineer proposes to use, their experience in mining, metallurgy and design of plant necessary to treat the ore. The engineer should be selected, perhaps after the client has suggested a joint venture with another group with complementary experience, on the basis of the experience and maturity of the people available and their knowledge of the particular geographical area in which the project is being developed. Remuneration of the engineer should not be a governing factor.

6. In the mining and metallurgical fields, most of the engineering companies are North American, although there are a few specialist european groups. Expertise tends to be concentrated in a few large groups with many years of experience in mineral/metal developments, and who have the necessary resources to develop very large projects in remote areas. The question of resources, both in the Head Office and availability for the field office, is of vital importance.

7. As a general rule, the Engineer's activities include the design, procurement, expediting, construction management functions and control of sub-contractors. Most of the manufacturing and site work will be sub-contracted on an international basis to contractors who have been pre-qualified by the Engineer and approved by the Client. However, it is not unknown for the Client to include in the Engineer's Contract a clause permitting direct employment of labour to carry out some or all of the site works. Similarly, the

Engineer's Contract with his sub-contractors should allow the
Engineer to take over the sub-contractor's work in the event of
non-performance. Because of the great problems of controlling
deliveries to remote overseas sites, considerable effort must be
devoted to inspection and expediting of all key items of plant and
equipment. This particular activity is rarely well executed, and it
is becoming more and more important as standards of workmanship and delivery dates are undoubtedly deteriorating throughout
the world.

8. Clearly, the development of ore-bodies is a specialist activity
and the Client must play a large part in determining the key
parameters of the project, a role which might be quite inappropriate
for other projects. For instance, the development of new port
facilities, dry docks or hydro-power schemes require skills which
only few clients will possess and the client may not even be able to
develop the scope of the project. In such cases, the client must
appoint an engineering consultant to advise on the proposed
scheme at an early stage.

9. While there is an increasing tendency for contractors, as
distinct from consultants, to be involved in the study stages of
such projects, it is as a general rule highly desirable for the client
to have completely impartial advice as to the best scheme, scope
of work and method of development. A possible exception to this
method of development could well occur if a particular contractor
had significant experience in the country concerned or with the
particular client.

10. Whichever type of organization is selected, the vital factors
are

   (a) the availability and experience of the personnel offered in
   the particular type of project being considered
   (b) the organization's experience in the particular geographical
   area.

It could be disastrous to select a consultant and/or contractor
with deficiencies in either of these areas. Irrespective of the type
of project, it is clearly of the utmost importance for both the
Client and Engineer to spend time and effort on understanding
national and local government statutory and guidance procedures.
In addition, in many areas of the world it is very important to
have a good knowledge of the anthropology and ethnic characteristics of the country/area being developed.

11. The most important additional points which need to be considered when there is a choice of sites for a potential development are

(a) industrial relations. It is a well known fact that in particular areas of the UK, and almost certainly in other countries, labour relations are worse than in others. If at all possible, clients should avoid areas with poor industrial relations records because of the disastrous effect that delays and over-manning in subsequent operations can have on the final return on projects. There are increasing signs that clients, particularly the more sophisticated, are paying much more attention to this critical area.

(b) labour adaptability. There is good evidence to show that there are many areas of the UK/world where the local inhabitants find it difficult to adapt to new industrial complexes and there is much to be said for siting new projects in areas of traditional, although perhaps changing, skills.

(c) the different fiscal and investment incentives available to international companies. These can be the vital factor in location of a new industrial facility; this is particularly true of automobile, petrochemical and oil companies' investments.

Financing the project

12. With the recent inflation of the world's major currencies, projects have increased in cost. Some twenty years ago, the £100 million project was large, today there are many in the £250 million to £1000 million category. Size brings new problems to an already difficult area. The capital cost is often too great for any one, even very large, company and projects must be carried out in joint ventures. This in itself can lead to management difficulties. Governments and public agencies can, however, develop projects which could not be supported by private companies. In turn, because of size, it is usually necessary to form consortia of engineers and/or contractors to carry out the work, which can lead to difficulties in bonding and guaranteeing: a topic of much debate over the past few years. This problem has been brought out almost solely because of size and high capital cost.

13. Financing of large projects leads to problems; there is no

doubt that the size of many new schemes has resulted in tremendous strains on the international financial markets — indeed, many new schemes have proved impossible to finance because of total cost and credit rating of the client, whether Government or private sector.

14. A particular example of the problems faced by the mining and metallurgical industries is that of completion guarantees. During the development of major mines in the 1960s it was not uncommon for the lending institutions to provide a completion guarantee to the project developer, sometimes in return for a share of the equity or for a commitment fee. As far as the lenders were concerned, this guarantee was underwritten by long term sales contracts, often with a floor price, to buyers of the highest possible repute. The revenue on these long term contracts was invariably sufficient to service and repay the debt. However, the increase in overall costs has resulted in, say, a 20% overrun on capital cost, or perhaps a year's delay to a project, leading to very substantial extra financial requirements — often £100 million to £200 million on completion guarantees. Lenders are not keen to guarantee overruns of this nature, particularly when the sanctity of contract is no longer regarded as inviolate. Particular problems have also occurred in many underdeveloped countries in Africa and South America where, often for external reasons, foreign exchange earnings are insufficient to service borrowings on completed projects. Obviously, under these circumstances, it is difficult, if not impossible, to finance new large projects with overseas borrowings in such areas.

Project risks

15. Size or large capital cost imposes great risks on all concerned, particularly on the client, who is much more exposed than the consultant or managing contractor, who can often predetermine and limit his exposure. There is no doubt that one major failure in a large capital project could almost bankrupt most of the private sector companies in this country. Certainly, the delays and overruns in some of the public sector projects seen in this country (for example Dungeness B Nuclear Power Station) could probably not have been carried by any private sector groups in the UK; in other words, shareholders funds would have been

eliminated. Consequently, failure can hardly be tolerated and it is necessary to have the most rigorous project overview by the client to determine corrective action on problems at an early stage or, if matters get out of control, often quite obvious to many, to abandon the scheme. At least one major mining venture in Africa was abandoned in 1975 with the total expenditure in the region of £100 million all written off by the developers; under the circumstances to carry on would have led to an unmitigated financial disaster for those concerned.

16. Some large or so-called 'jumbo' projects will be developed if only because their owners, whether government and/or private sector companies, perceive a significant gain in large-scale developments. However, much of the more recent comment by both industry and Government on the need to be involved in large projects should be treated with caution and perhaps even scepticism. Large projects are much more difficult to achieve and complete and therefore profit, certainly for the client, is more difficult to earn. Given a choice, most groups would prefer to deal with ten £100 million projects than one £1000 million scheme. Caution is undoubtedly the watchword in this area and only clients and organizations experienced in large schemes should contemplate the largest of schemes.

Developing the project

17. There are certain essential guidelines which must be followed in developing any project which assume an even greater significance in large capital projects.

(a) The client's team must be of the highest possible calibre. It must be involved continuously through inception, evaluation and execution, and must at all times be quite independent of the engineer.

(b) Sufficient finance must be committed by the owner to thoroughly investigate the project. This finance, which is at high risk, can in some schemes amount to over 5% of the total capital cost and, therefore, if the owner cannot contemplate writing off this sum, he should not even commence the study!

(c) Financial analysts, tax experts, accountants and engineers must all be included in the client's group. Far too often the

client is represented by one professional discipline, which is
rarely a satisfactory solution.

18. Ideally, a large project should be developed in stages, the first
being the preparation of a feasibility study. This study, even though
preliminary, should be comprehensive and should examine all
aspects of a project and develop adequate or alternative solutions
for all areas. Since, after this stage, large sums of money will be
spent, it is vital that this study be carried out well and in depth. It
should be reviewed by a group of experts with external advice to
determine whether the work has been well done and the conclusions
properly drawn. There is little doubt that many so-called jumbo
projects would not pass this preliminary stage if the study was
properly executed.

19. On successful completion of the preliminary study, it is then
necessary to complete the detailed planning, preliminary engineering and preparation of budget estimates, prior to financing. Of
particular importance at this stage will be the preparation of the
capital cost estimates. All too often this area is inadequately
handled, usually because of one or a combination of the following
circumstances:

(a) inadequate definition of scope of plant, equipment and
facilities; it is vital to prepare proper flow diagrams, layout
drawings and plant and equipment lists
(b) insufficient attention being paid to quantities of materials
required and the cost of purchasing or obtaining these materials
in the project location. Often data is extrapolated from one
country to another with insufficient attention being paid to
differences in productivity levels, costs of materials, etc.
(c) the programme of site work being too optimistic.

The client must review all the above for both optimism and pessi-
mism before proceeding to financing.

20. An additional point often missed by the client, which is
particularly important in remote locations, is that it always pays
to spend money on installing plant and equipment which is robust,
sometimes over-engineered and highly reliable, even if the operating
costs are higher. Frequently enforced plant shutdowns, due to
equipment breakdown and a lack of spares, lead to serious loss of
availability and/or profits. Such a concept can present difficulties

when competitive bidding is the key word — the client must be quite clear in guiding the engineer in this area.

Conditions of contract

21. There are basically two types of construction contracts: the fixed price or so-called lump sum turnkey system, and the reimbursable costs plus management fee, where the client pays all costs plus an agreed fee. Although there are many variations of the two systems, the distinction between them is quite clear and they have different consequences for the client. On the whole it is better to contract the supply of plant and equipment on a lump sum basis, albeit with an escalation formula, and to carry out site work on a cost plus fixed fee basis. With proper control, this system allows the client the right to carry out essential changes without the risk of undue financial penalty and makes it much easier for the contractor to tender without fear of great financial exposure. Obviously, with the cost plus fee system, the client has to have a rather more sophisticated control system.

22. There are many variations in the terms and conditions for major manufacturing or construction contracts, and much has been said about the advantages and disadvantages of the different versions. The larger or more remote the project, the more the client must relieve the contractor of unfair exposure. Otherwise, the result will almost certainly be over-pricing of the project. In other words, the Terms and Conditions of Contract must be equitable to the contractor and the client with the risks residing in the appropriate quarter.

6. Project management

W. J. Ryder, FIMechE, FIPlantE, FInstPet, MBIM, MASME*
J.H. Mercer, BSc, FIChemE[†]

Introduction

1. Management of large capital projects usually involves a number of different parties, principally the client and the main contractor, but with many other experts, vendors, sub-contractors and specialists, often with very divergent views. To mould this total effort so that a contract is executed efficiently has resulted in the rise of specialist contractors who can direct and control such a complicated and interdependent force.

2. The Authors have spent a major part of their working lives in the Process Plant Contracting Industry and the Paper is based on much of the experience gained during this time. However, the principles apply equally to contracts in the civil or mechanical engineering fields.

Choosing your contractor

3. The UK has a wide choice of contractors and in the Greater London Area there are about thirty of different specialities and sizes in the Process Plant field alone. It is, therefore, important that a client chooses a contractor most suited to the particular project in hand, considering both the size and type of project envisaged. Very large contracts can only be handled by the larger contractors. Conversely, a very small contract can get lost in a large contractor's office. Each contractor will have different ideas about the amount of work he undertakes by direct employment

*Managing Director, Woodall Duckham Limited, Crawley
†Director, Woodall Duckham Limited, Crawley

of personnel and how much he sub-contracts to others. The latter method may require more control as the work is executed in several locations.

4. A different type of control will be required where much of the design and supply is sub-contracted into discrete packages. This may add to the programme time in order that the principal Contractor and Client reach a firm stage of definition for the total package, so that any variations or additional work on each discrete package are kept to a minimum.

5. Most contractors appreciate clients who carry out a pre-qualification visit, during which the broad outline of a particular project is given by the client, and the contractor states the relevant experience that he has established in the particular field in question. There are numerous factors which are often discussed, such as

(a) contractor's forward load
(b) possible candidates for the project manager's position
(c) company's previous experience in the given field and the availability of specialists needed for a particular contract
(d) the company's systems associated with project control
(e) the broad outline of contract conditions.

The pre-qualification information must of course be treated as strictly confidential. The advantage to contractors is that it results in only a limited number of companies being invited to bid against the client's tender documents. This saves a great deal of bidding effort and associated cost when taken over the whole of the contracting scene.

6. Although in general all contractors operate the project management system, there are many variations in the systems used; each company adapts the general system to suit the type of work and the clients from whom they generally obtain business. A typical project organization is shown in Appendix 1 (see page 60).

Choosing the type of contractual conditions

7. The determining factor for the type of contractual conditions is the state of knowledge that the client has for the particular job in mind. If the requirement can be precisely defined, i.e. 1000 t/d nitric acid plant or a 1200 t/d ethylene plant from stated raw

	Head Office costs	Material supply	Civil and building/ construction	Commissioning
Good definition	Fixed price			
	Firm price with escalation			
Reasonable definition	Firm price with escalation		Reimbursable	
	Firm price	Reimbursable	bonus/penalty target price – sharing overs/unders guaranteed maximum cost	
Poor initial definition	After design freeze	Reimbursable		
	Reimbursable			

Fig. 1. Examples of main types of contract

materials, then there are several capable contractors who could give a firm price offer for Head Office services, materials supply and possibly construction. At the other end of the scale where the client has a new process which has only been developed to the laboratory or pilot plant stage, where the requirement is for process development and scale up to a production capacity, then a fully reimbursable contract may be equitable to both parties. However, after the flow diagram and layout have been 'frozen' it may be possible for the contractor to firm up certain of the costs of the job.

8. In between these two extremes, there are many variables which have already been presented in Paper 4 and will be discussed in Papers 7 and 8. The main types of contractual conditions are given in Fig. 1.

Running the contract

9. As an early step to ensuring a successful project it is essential that both the Client and the Contractor appoint a person whose sole responsibility is to ensure that the contract is completed, if possible, to everyone's satisfaction. In both organizations therefore this key role is given to a project manager who has within his control a selected team who will remain with the job for its

duration and mobilize from either the Client's or the Contractor's organization the people necessary to give information, provide approvals and assign the manpower to carry out the work. The essential feature is that there is a single channel of communication between Client and Contractor and each project manager has the facility to obtain fast reactions from his own organization.

The five phases of a contract

10. Any project may be divided into the following:

(a) tender submission, negotiation and signature of contract
(b) the design phase — preparation of engineering specifications, drawings and requisitions
(c) the procurement phase — material purchasing, expediting and delivery
(d) the construction phase — civil, building, mechanical and electrical construction
(e) the commissioning and operation phase.

Each of these phases are different, take place at a different location and require different skills to control. Phases (a) and (b) are mainly based at the Contractor's Head Office, phase (c) starts in the Contractor's Head Office but then extends over many manufacturers' works in the UK and (possibly) overseas, and phases (d) and (e) take place on the job site which is usually far removed from the Contractor's Head Office.

11. The organization of the project as a whole and the control by the Project Manager must be such that all of these phases, in spite of their differences, become an integrated whole and the inter dependence of the one phase on another can be recognized. He must see that any upsets in one phase have the minimum effect on the subsequent phases. To do this there must be

(a) good control of the technical content
(b) good control of expenditure
(c) good control of calendar time
(d) good control of human resources.

These are the essence of project management.

12. Papers 9 and 10 deal with human resources but we will look at the other points of control.

Controlling the design phase

13. Everything that goes into a job has to be designed, and a purchase order has to be placed, for every item or service required. The whole basis of time and money control therefore starts in the Design Department. In addition, under the Health and Safety at Work Act, designers have a clear-cut responsibility to ensure that the plant which they offer is safe to be operated.

14. It is therefore incumbent upon Design Departments to have a rigorous system whereby the various grades of personnel, from draughtsmen through professional engineers to managers and directors of engineering, have their responsibilities defined and various 'stop and check' points are undertaken during the design phase of the job. These generally consist of review meetings, where flow diagrams, layouts, calculations, specifications of major items,

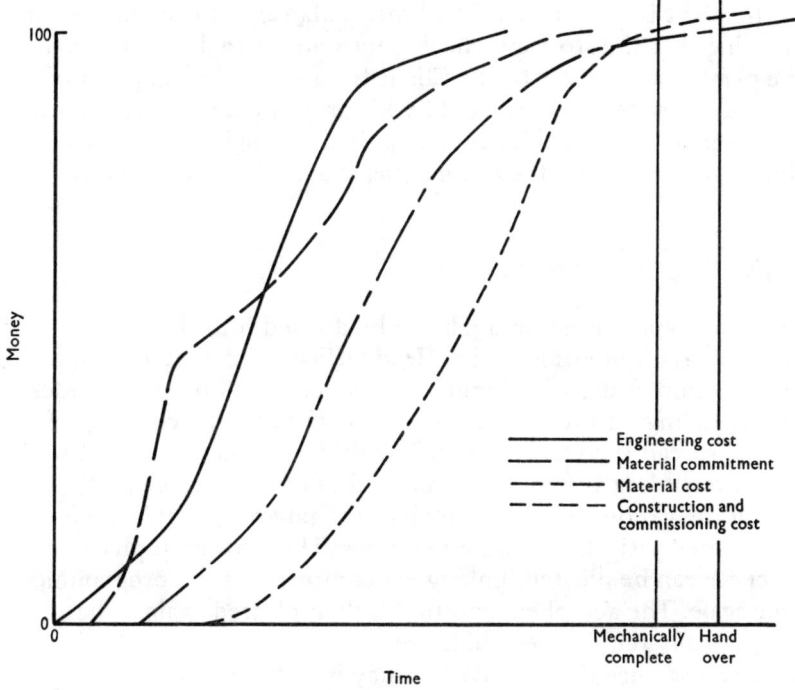

Fig. 2. Cash flow

the hazard analysis, etc., are subjected to critical examination by a person or committee not directly involved in the project. These stop and check points may also involve the Client, especially where customer standards are being worked to or where the operation and maintenance of the plant is being considered.

15. Good records must be kept of the 'design considerations' and of the 'review meetings' so that it is possible to look back when the plant is being commissioned to see that the design requirements are being fulfilled. In the circumstances where a main Contractor is required to sub-contract part of the design there is still a great need for direct supervision in order that the technical content of the sub-contractor's work is subject to the same scrutiny as in-house work.

16. The control of technical matters does not end in Head Office supervision. Items of equipment fabricated in maker's works must be inspected to the required standards; similarly, the construction effort must also be subjected to quality control so that the integrity of the plant is maintained. The Project Manager is responsible for ensuring that the stop and check points are carried out and that the plant is acceptable to the Client from the technical point of view. To this end he is assisted by his project engineering team who have a great responsibility in seeing that the engineering co-ordination is good and that the engineering standards are adhered to.

Controlling expenditure

17. The expenditure on a job can be divided into the same headings as given in Fig. 1, i.e. Head Office costs, material supply, construction costs, and commissioning costs, and it occurs under these headings at different times during the life of a contract. A typical spread is given in Fig. 2. The Project Manager must also have a recording system and a method of control to see that, during the various phases of the job, the rate of expenditure is in accordance with the programmed rates. Thus, for each phase an 'S' curve can be plotted, linking expenditure and the programme time scale. The actual expenditure is then plotted against the theoretically required expenditure.

18. The essence of control of money is to have

(a) a reliable base — a review of the estimate with the necessary revisions to be made when the design is frozen

(b) a control point which authorizes expenditure prior to the commitment being made

(c) the 'cost to complete' — a continuous reassessment/prediction/re-estimate of each phase at regular intervals

(d) cash flow — a statement of money committed, and the anticipated and actual cash flow

(e) escalation — a statement of the anticipated and actual escalation.

Time and money are synonymous. The graphic representation of the expected expenditure of money against the actual is a very useful guide as to the progress of the job and whether or not it will be completed on time. In simple terms, if the money is not spent quickly enough, the job will be late.

19. In any plant there are certain areas where it is difficult to estimate accurately with the information available at the tender stage. The control of expenditure for these particular sections is equally difficult. Thus, special methods have to be adopted to keep under control such items as

(a) the purchase of bulk materials, i.e. pipework, electric cable, instrument fittings, etc.

(b) the cost of a development item where running tests in the maker's Works may involve redesigning and remaking parts which fail under test

(c) construction costs where weather, industrial relations, delays in shipping or port clearance, loss in transit, etc., have a significant effect upon final cost.

The charts in Fig. 3 show the effect of various upsets that can occur in the life of a project.

Role of the Project Manager in cost control

The base

20. The Project Manager has to decide whether the estimate prepared in the tender stage is in sufficient detail to become the Cost Control Document in the contract stage. If the estimate has been prepared in sufficient detail it is probably suitable. However, if the estimate has been prepared in a short space of time it may

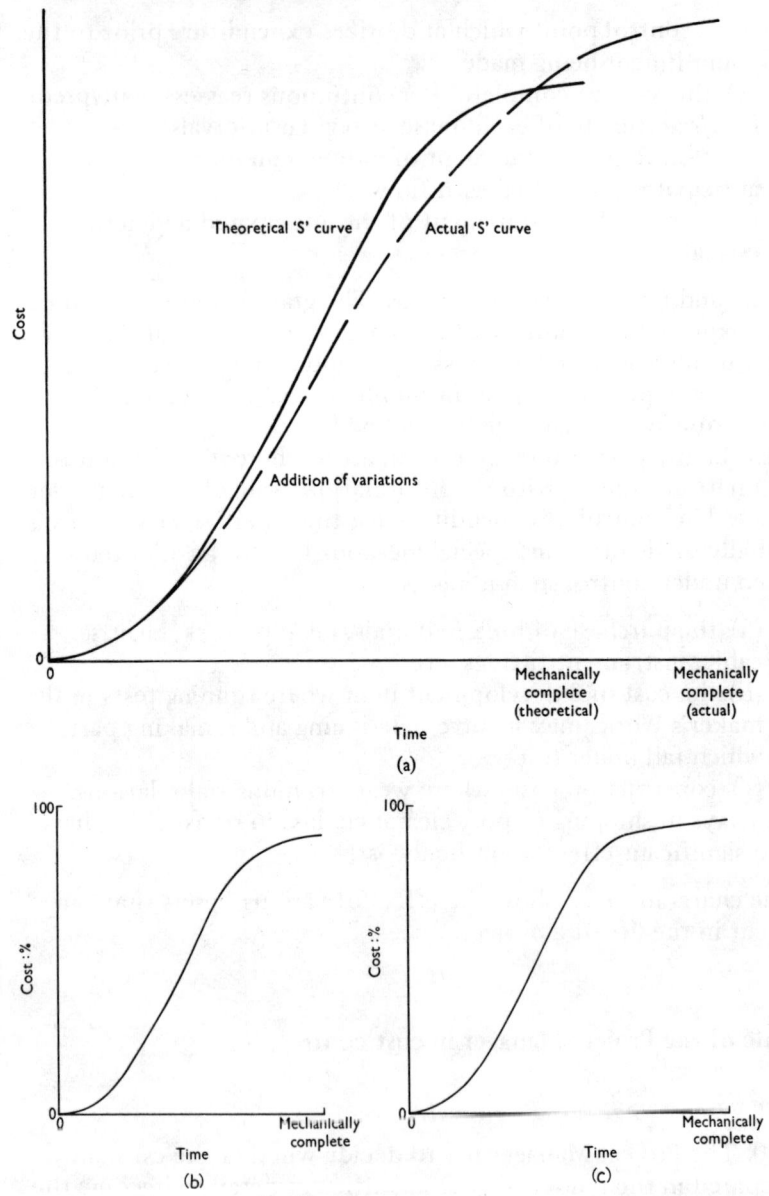

Fig. 3(a) Cash flow; (b) Total cash flow 'on time' contract; (c) Total cash flow 'late' contract

be necessary to prepare a more detailed estimate which can be used more adequately as the Cost Control Document. This re-estimation can possibly be left until the design freeze stage has been reached.

The control point

21. The control point is generally the Cost Control Officer who forms part of the project team. He has certain rules laid down, generally working on the 'by exception' principle, i.e. all things being equal, if quotations received for an item are equal to or less than the estimated cost, then it may be purchased, but if it is over the estimated cost it requires initially the Project Manager's approval as this will result in a reduction in his contingency sum. The Project Manager may also lay down the rule that he requires to authorize

(a) all purchase requisitions over a given value, and
(b) purchase requisitions for certain key items, which either have a significant effect on the technical working of the plant or the execution of the programme.

More senior management approval is also desirable where orders in excess of a prescribed sum are to be placed, or sizeable overspends are required.

22. The Project Manager again becomes involved when there is a difference of opinion as to which supplier should be selected: where the technical people favour Vendor A and the procurement people favour Vendor B. The Project Manager, having the overall view of the job, can make up his mind whether to adopt, for example, technical excellence at a dearer price, or shorter delivery at an enhanced price, or a price within the estimate that does not have such a good technical specification, but is adequate.

Cost to complete

23. The Project Manager must hold, at regular intervals, a detailed review of the costs incurred to date, and the trends that are appearing, both within the job and nationally or internationally, as far as escalation and hardening of the markets are concerned, so that a view can be taken in predicting the anticipated final cost

of the project. This will involve a great deal of forward projection, an appreciation of the detailed programme and a judgement as to the likely difficulties that may occur and that will need some financial provision to overcome in the future. This function can only be developed as a result of experience. Obviously if a company has a 'standard' line of business it becomes easier, but if a company is involved in 'one-off' jobs, the skill of the Project Manager and his team is all the more important.

Escalation and cash flow

24. With the escalation rates that have been experienced over the last few years coupled with the high interest rates, the control and flow of the money is of paramount importance. The terms of payment laid down in the contractual conditions and the escalation formulae (if the contract is on an escalatable basis) require a great deal of expertise to apply and a promptness in submission and payment of bills to ensure that the cash flow situation is controlled or improved upon. If it deteriorates then much of the potential profit can be quickly eroded.

Controlling the programme

25. As for cost control, one has the same basic requirements as far as the programme control is concerned.

A reliable base

26. Generally speaking contractors have found that it is worthwhile at the tender stage developing an overall network with a limited number of activities to ensure that the logic of the proposed method of carrying out the work is basically sound and gives the minimum contract period. The interdependencies resulting from this are then clearly shown with the critical path and the immediate sub-critical paths identified. As a result, on receipt of the contract, the Project Manager and Project Programmer can then develop the Tender Critical Path Network (CPN) to a full Contract Network.

The control point

27. The Programmer is the centre of the time expenditure network. He is dependent upon many other people, perhaps in different locations, reporting against completion of activities or giving 'will complete by' dates. In monitoring and controlling cost there are documents, such as requisitions, that have to pass through the control point. There is no similar document passing through the control point related to time expenditure. Consequently time control is more difficult and very much dependent upon the Project Programmer continuously monitoring activities and making his own judgement as to the completion date of many activities, and hence to the completion date of the project.

Continuous record of programme activities achieved

28. *Design phase.* The sale of technical man-hours is a major part of the Contractor's business and the control of the use of these man-hours is essential. Quite sophisticated systems may be involved in the estimating, recording and predicting of man-hours required, hence the programme time necessary for the completion of individual drawings and specifications. Typically the key dates for completion of such documents are extended usually on a computer print-out, so that each section of the Design Office may have a detailed list of drawings and specifications to be prepared, the calendar time and number of man-hours allocated for each. The Project Engineer and the Project Programmer must ensure that the interdependencies are highlighted, that all the sections fit together in the overall scheme, and that there is communication where necessary between different design sections.

29. *Procurement stage.* For this stage the overall programme will be extended to a list of every item of equipment, material or service that has to be purchased and delivered to the job site. For each item the programmed dates on which it must be ordered and delivered are stated. When expeditors visit vendors' works, anticipated delivery dates can be determined and compared with those planned. If necessary remedial action can be taken where possible to safeguard the end point.

30. *Construction phase.* In the present climate the programming

and control of time is without doubt the most difficult of all. The construction function is at the end of the queue and suffers the results of lateness by all other parties. It also has to cope with the effects of inclement weather and above all the possible effects of industrial disputes, not only on the site itself but in such places as the vendor's works, the docks, and other transport areas. The construction programming therefore must be of a very high order to be able to maintain the optimum labour force in productive work and yet complete each section of the job in the correct order, thus enabling the Commissioning/Operating Department to carry out the pre-commissioning tests and set the plant to work. Each foreman and supervisor will require a work list which will tell him that the material is available and that his gang should complete the work in a given time. So construction programming is a detailed affair but all the detail must also fit in with key dates in the overall programme to ensure the correct flow of drawings and materials from one phase to another.

A continuous reassessment

31. The priority order for review and reassessment of the project end date is determined by the critical path and the next few sub-critical paths. However, descriptive reviews of the detailed component programmes must be undertaken at regular intervals. This will necessitate not only reviews at Head Office of the information received but also visits to vendors' works and sites.

Variations

32. Variations in the scope introduced during the course of the job have a significant effect on the programme. A series of small variations, each in themselves not considered significant, together can add up to a major cumulative delay which will not be apparent from a simple addition of the number of small variations. Clients and contractors should if possible agree to some simple rule, such as 'variations can be accepted up to the freeze stage'. Thereafter the possible effect of changes may have a serious effect upon the programme with all the attendant costs.

Conclusion

33. Completing a job on programme within the estimate and achieving the guaranteed performance to the satisfaction of both Client and Contractor is the Project Manager's remit. As everyone appreciates this is a very difficult task and one that can only be achieved if both Client and Contractor are intent upon working together to attain this objective. Both sides must therefore make the right decisions in choosing the type of contract, the right contractor and his project management and systems, and also in ensuring that the general outlook and philosophy of the companies and the individuals in them are compatible. If they turn out to be incompatible, contracting no longer is fun but an extended period of torture.

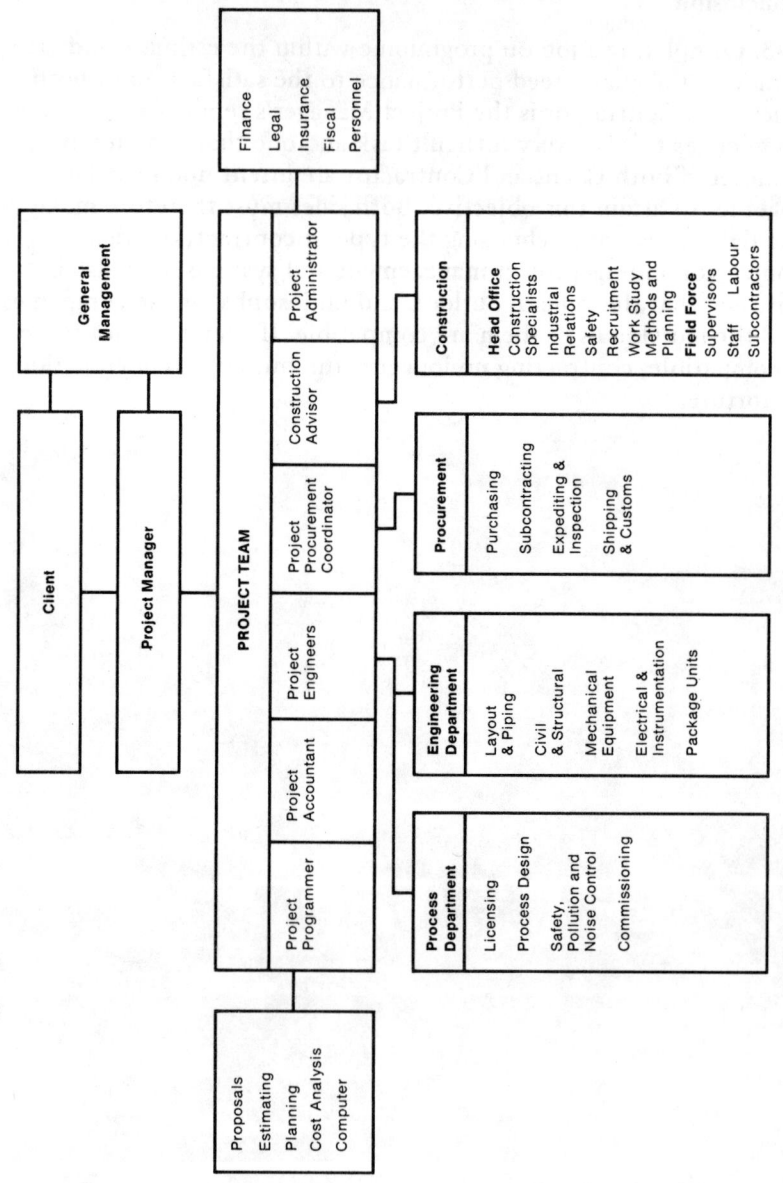

Appendix 1. Project organization chart

Discussion on Papers 5 and 6

Dr N.M.L. Barnes *(Martin Barnes & Partners)*

1. There are lessons to be learned from the whole range of engineering projects, but it is my view that some of the techniques of project management mentioned in the papers should be modified in the civil engineering context.

2. On the selection of the form of contract, I agree that a criterion for choosing one of the less conventional forms of contract is lack of firmness of design. I suggest that another reason for selecting a less conventional form of contract, such as a target cost or reimbursable arrangement, is when the client, through his consulting engineer, wishes to exert more than the conventional amount of management control during construction.

3. One of the shortcomings of standard forms of contract is their assumption that the employer intends to leave the contractor alone during construction. On many large projects this is untrue, as the employer needs to exert control in order to maintain completion dates in uncertain circumstances.

4. Procedures for controlling time and cost should be subjected to the test of whether they work on a job which changes significantly between tender and final account. To pass this test, more forward-looking procedures have to be used than the simple comparison between a detailed estimate and costs to date. Because of the intrinsic uncertainty of civil engineering work, the detailed budget drawn from the original estimate is not a good pointer to what individual costs ought to be. Using a 'cost to date' type of cost control system, something in the order of one in six of cost centres will be showing a 10% loss at any one time — simply because of the uncertainty of estimating. These losses are spurious: uncon-

nected with the ability of management on the site or with the success with which the work is handled.

5. If a comparison of actual cost with estimate is used as the basis of cost control at a detailed level, management effort often may be directed at work which is already being done as cheaply as possible in the circumstances. A more forward-looking style of control is needed. This is best instigated by bringing together the time and cost control.

6. Managing projects is essentially managing activities which both generate cost and absorb time. Managers should be equipped with the means of costing out and timing out their proposals as to how to do work. This can be an integrated and essentially forward-looking function.

7. In paragraph 27 of Paper 6 the Author suggests that planning engineers have a special difficulty in that they do not have paper work to indicate to them the amount of time that has passed. This is compared with cost controllers who have access to information about cost to date. In my view this is an advantage for the planner. If he wants to know how much time has been absorbed in doing the work to date he can look at a calendar. It is my experience with cost control systems used on many large projects that cost controllers spend too much time looking at the paper work to try to form a view of costs to date and not enough time considering how to reduce or minimize the cost of the remaining work.

8. Some of the large projects in the Middle East and in other parts of the world now use more forward-looking aids to project control. This is encouraged by recent computer developments and by the development of cost modelling techniques.[1]

9. The current style of project management is characterized by

(a) detailed cost control comparing actual with estimate
(b) detailed networks prepared separately from cost control and management
(c) control discipline exerted through cost and time variances.

The future style of project management will be characterized by

(a) cost control emphasizing forecasts to completion
(b) time and cost control merged
(c) information designed to support decision making

(d) control based on co-operation and delegation rather than hierarchical discipline.

Mr H.J.C. Faulkner *(Thames Water Authority)*

10. Planning programming and time control is a major weakness in project management in this country. Dr Barnes and the other speakers seem to accept, in the organizations with which they are in contact, the importance of time control. I wish there were more organizations like these. One of the major problems in Britain is the failure of management to appreciate the importance of time and planning of a project, from its earliest stages, through the design, manufacturing, to site construction. There are problems in planning; it requires a rigorous mental discipline to try to forecast and look four or five years ahead. This is not very popular, particularly on the part of top management.

11. In the planning and programming process the importance of top management itself being completely committed and involved in the planning process is essential. It is no use having some sort of organization where a planning engineer is trying to operate at the lower level, usually in a vacuum, where no one else is taking any notice of him and the people above him are not particularly interested.

12. The major contractor is well equipped to plan, to involve human and money resources in sophisticated techniques, critical path planning and so on. However, the main contractor is dependent upon the smaller contractor who, in his turn, is dependent upon his sub-contractors.

13. How does one control smaller contractors or sub-contractors who are involved in quite a large project? Frequently, once a contract has been let to a contractor, the responsibility for the timekeeping of the job falls to him and not to the consultant or the main contractor. I feel that the consultant should involve himself in the supply, planning processes, and functioning of his suppliers and his smaller contractors, to ensure that his critical equipment will be delivered on site on time. A contractor or consultant ought to be able to assist a small firm which has not the resources to plan adequately, and to help out in the project as a whole.

14. If more effort and thought could be given as to how this function could be successfully implemented, overall project management would benefit to a great extent.

Mr D.W. Berry *(Howard Humphreys & Sons)*

15. Regarding the 'S' curves shown in Paper 6, could the Authors give any idea of the confidence bands on these curves? A commonly used equation for the 'S' curve is $y = \sin^2 x$. However, I have never seen confidence bands equated relative to that formula.

16. The Authors' 'S' curves did not appear to take into account the case where a down payment is made. Should the 'S' curves not start above the origin in such cases?

17. Another easy-to-use approximation which I often adopt for projects up to three years' duration is to assume that 10% of the contract expenditure occurs in the first four and the last four months of the contract period, with the remaining 80% spread over the central period. This corresponds fairly well with the formula quoted above.

Mr M. Milne

18. Reference was made in paragraphs 16, 18 and 19 of Paper 5 to the problems of the large contract and, in particular, to the fact that any wise company would rather have ten contracts of £100 million than one contract of £1000 million.

19. Many enterprises which are not international but national run through the life of several administrations. The commitment is entered into by one administration which is not necessarily accepted by those which follow. For example, the TSR2 had a large commitment of expenditure and was then cancelled. Then there was the later coincidence of three major different types of projects, all coming to peak spending at about the same time: Maplin, the Channel Tunnel and Concorde. Of the three, it was decided to proceed with Concorde, despite its certain losses, because of the political and other commitments which had been made.

20. A large organization in a developing country can call on

export guarantees and so on, to protect the engineer, the client or contractor against insolvency. However, a large international organization, dependent on the backing of a single government, has problems when entering into a large commitment of time, money and expertise with, under it, a whole range of other organizations, looking to it for engineering support, financial commitment and management expertise. How does it cope with the situation of government change of policy of a very large magnitude — which must distort both the finances in the current sense and the expectations of a whole range of enterprises subordinate to the main management organization?

Mr M.A.C. Luetchford *(BP Chemicals Limited)*

21. Much of the development work and financial commitment for North Sea oil was made at an early stage, when client companies did not know the future tax position. It was not until 1973 or 1974 that taxes in the form of PRT were established. While in the event this is not an unreasonable tax in its own right, the original commitment to North Sea oil was made as an act of faith, since it was not possible to evaluate the full profitability of projects until the tax situation was known.

22. It is difficult for any client to be able to establish in advance what his return on a particular project will be, since changes can be made even by relatively stable governments which can affect the return of capital long after finances have been committed.

23. At present, in the chemical field in particular, there is an over supply situation, and the establishment of chemical prices over the next 3—5 years is difficult. Thus, to some extent it also becomes an act of faith for a large client to commit funds to major chemical projects.

Mr R.W. Postlethwaite *(Partner, Peter Fraenkel & Partners)*

24. It seems nowadays necessary to take the potential industrial relations aspects into account during the design of projects, as it is clear that there are certain parts of this country and elsewhere where particular methods of construction are almost bound to

lead to strife. Thus, in order to ensure the timely completion of a project it is necessary to consider whether certain construction methods might be entirely eliminated and replaced by those that will be less likely to lead to labour problems.

25. Have the Authors of Papers 5 and 6 considered altering designs in order to obviate and eliminate possible industrial relations aspects that they have foreseen on their projects?

Mr D.A. Meyers *(Freeman Fox and Partners)*

26. Large capital projects are growing in complexity and the function of management is becoming increasingly more difficult to perform effectively. The consequences of errors of judgement, slow response to a deteriorating situation, or lack of a complete and proper overview of what is going on, can be serious on a large project; yet the task of coping with all this becomes overwhelming if left to manual resources alone. There is simply too much information to be disseminated and analysed for this to be done effectively without some form of computational aid.

27. However, many experienced and capable project managers have suffered under computer systems for project management that have added to their problems rather than relieved them. Since the first rudimentary systems were thrust upon an innocent fraternity of project managers by a starry-eyed group of computer 'boffins', much fur has flown. However, as a result some sort of commonsense approach has gradually emerged and progress is being made to improve the effectiveness of computer-based management systems. Some useful improvements in techniques have evolved over the past few years — with the active participation of engineers, merchant bankers and others who are necessarily and properly involved in the complex, total process of project management — and these techniques are being incorporated in systems which are far more versatile and realistic to handle than earlier versions of a decade or so ago.

28. The INTERNET system (referred to in discussion on Papers 1—4, paragraph 34) has been in use over a number of years and an extended version of this system is currently under development with the support of the National Research Development Corporation. The scope of this system extends from financial modelling

through other parts of the wider spectrum of project management embracing, and if necessary integrating, the areas of planning and procurement. All these are vital areas within the domain of project management and merit serious attention. It is therefore important that project managers should not disregard the powerful and flexible tools that are being developed to assist them with these tasks.

Mr S.E.J. Beare *(Director, Stuart Beare Associates Ltd)*

29. Management is part of engineering, engineering is part of management; the two philosophies are separate yet complementary, and must somehow be operated more efficiently together.

30. In varying organizational structures, how many project managers have the absolute authority to act in such a way as to benefit the overall project objective? The authority and responsibility delegated to the appointment is usually insufficient. Consequently, it is only rarely that this vital position is filled with a person having the required expertise and stature.

31. The experience of changing philosophies should not be missed by those persons with an engineering background who find themselves thrust into management positions. It seems wrong that after gaining years of experience, engineers, because of their status within a practice, are then expected to be as good a manager as an engineer. This is not to say that it is impossible for this to happen. Indeed, it could be more the rule than the exception if the in depth training of the profession improved.

32. The appreciation of the client's problems is most important. The full involvement of the client and his representative with the project team is essential to ensure that the true effect of every decision is known in advance.

Mr T.D. Kershaw *(Consultant)*

33. As a project develops, the centre of activity moves from design office to site but all too frequently the design office is reluctant to recognize that the decision making focal point has moved to the site and that delays can result. The project manager

must have adequate authority to ensure that this does not happen. His presence on site, when activity is concentrated there, may be essential and if he cannot be there the delegation of authority to the senior man on site must be the maximum possible, even to the extent of taking some risk. If such delegation is not possible, either the management structure is wrong or the senior site representative is inadequate. In either case, the project will suffer.

34. With reference to the construction of hydro-electric projects overseas, there is considerable benefit to be obtained by consultant and/or contractor if they can staff their field organization with the senior man being a civil engineer and the deputy an electrical/mechanical engineer. When the construction phase progresses from its emphasis on civil work to electrical and mechanical installation, the civil engineer can be released and the deputy move into his position. This method has been tried successfully despite the difficulty of finding a suitable electrical/mechanical engineer.

35. The clear thinking and basic principles outlined in Paper 5 (particularly in paragraphs 3, 17 and 22) could be applied, after adaptation if necessary, to projects having the traditional employer/consultant/contractor relationship.

36. On the Tarbela Dam Project, which in 1968 was the largest construction contract ever awarded, the contractor's management team, which effectively carried out the works, felt the strains imposed by its scale and I feel that the size of the contract was at the upper limit for efficient management.

37. A section manager in the field had to be of agent capability, but contractual, personnel and union negotiations, planning and contract management, and so on, had to be centralized functions, under the Site Agent/Project Manager, so that one overall policy was adhered to. This effectively prevented the section managers, who were also under the general control of the Works Superintendent, from fully utilizing all their abilities without prior reference to others and resulted in some personal frustration at times.

38. Therefore, the obvious initial attraction to giant sized projects should be tempered with caution until a satisfactory answer to their management problems has been found. The projects could be divided into a number of smaller contracts, but this would create a different sort of management problem by giving the engineer or client organization a heavier management load, which few are equipped or experienced to carry.

Mr Frame

39. I have been involved in some civil engineering projects in the UK and I think it is dangerous to generalize and to say that civil engineering projects have a propensity to change. If this happens, it is often because the contracts are awarded before the design is properly advanced. It is up to the client to decide how he wishes to proceed. If he wants a good job, he should not place a contract until the design is sufficiently advanced.

40. Dr Barnes is clearly interested in systems. We have all seen systems come and go in project management, but they are no substitute for management. One must have a clearly defined method of carrying out a project and use the systems to help one to manage.

41. With reference to Mr Faulkner's comments, good clients become interested in time and cost control right from the beginning. Here, at the risk of being a little 'private sector versus public sector', it is perhaps worth noting that in the private sector in the UK the control of costs on projects is rather better than it is in the public sector. One of the reasons for this is that if some of the large capital costs were allowed to overrun, as can be seen in the public sector, in the private sector there would be a large number of bankruptcies. This makes private companies think very carefully and put their best people on the projects. I can think of many public sector projects in this country in the last five years which would have bankrupted the largest companies (in this country).

42. That is a good discipline for top management, whether public or private sector.

43. We are involved in the North Sea — not in quite as big a way as BP. But the PRT mentioned by Mr Leutchford, is a perfectly reasonable arrangement. There is no oil company which has developed oil in the North Sea that has not made handsome profit out of it. (In our own case, the pay-back on North Sea investment is faster than any mining development.)

44. Some of the oil companies undoubtedly took a major gamble in the North Sea. It is a gamble which, representing a prudent mining company, I would not have taken. If it had not been for the Arabs' increase in the price of oil, every company investing in the North Sea would have lost a significant sum of

money. It was a gamble, not an act of faith. If people gamble, they have to be prepared for the consequences. Certainly the writing was on the wall. We could see it ourselves, as a minor oil company. If it had not been for the increase in the price of oil, no act of faith in going ahead would have prevented a disastrous situation.

45. Investment in chemicals has nothing to do with the organization of large projects; it is how one views the world economy. Every industrialist has to consider this at the present time, whether he is big or small.

46. I shall try to answer Mr Milne's question (paragraph 20) in general terms rather than in specific RTZ terms. Luckily, I do not have to work for governments; we can decide where we do projects.

47. One of the great problems in working, whether one is a consultant or contractor for government, is continuity of government policy. I believe this can be done only on the basis of a project which is thoroughly worked out, thoroughly studied, evaluated and agreed. This was one of the big disadvantages of the Channel Tunnel project, with which RTZ was associated. The work was well carried out by the engineers, contractors and project managers. It fell down because the government could not stick with the going, mainly because a nationalized body made a mess of their particular part of it. But, on the other hand, it was carefully controlled, it was developed in a particular way by us as project managers in that the agreements were drawn up so that it could be aborted at appropriate stages, and the people who were involved were looked after (I mean companies and organizations). This is the only way for it to be done.

48. Concorde was an absolute disaster — the way the contracts were drawn up — and we know the story of Maplin.

49. I do not think you can rely on continuity of Government policy. Agreements must be drawn up in case they change their views, and one must work for governments in such a way as to be able to get out when changes of policies come, on terms which suit the contractor or engineer — in other words terms which are advantageous to them — and governments must agree them before one goes ahead.

50. This is the way we work overseas when we work with governments. It works with unsophisticated governments. If it

works with them, I do not see why it should not work with sophisticated governments — if there is such a thing!

51. Answering Mr Postlethwaite from my own and the company's
experience, luckily we can decide whether we go ahead with a
project in Britain, or Australia, or Canada, for business reasons. We
will take into account a very careful assessment of how long we
think it will take us to build a particular plant and how we will be
able to operate it. These will be major factors in our conclusions.
We have undoubtedly done things abroad which could have been
done in Britain, because we were concerned about time to build
and industrial relations. Everybody, with any real commonsense,
tries very hard to decrease labour content on site. We certainly try
to do this. I have tried to persuade a particular nationalized body
to do this in other areas. It is a very important factor, but there is
no doubt that to carry out a major capital project in Britain is
harder than in almost any other place in the world.

52. Look at British Steel's experience in major steel complexes.
I believe they completed a plant in South America in twenty-nine
months within budget, working with Davy International. They are
not able to do the same thing in Britain within two years of that
time, in a so-called 'sophisticated' country. This is a factor which
people should take into consideration in investment in the UK —
we certainly do.

Mr Mercer

53. I do not think that one can draw the distinction between
the current style and the future style of project management.
Time is money and money is time and the project team must not
be separated into the different areas but must work together.

54. The requirements are firstly that there is a water-tight system
which gives the actual expenditure to date and the actual amount
of work done to date and secondly that the historical data is
collected in such a way that when plotted against time, trends are
established. The experience of the project team — project managers,
planners and cost controllers and others — can then be concentrated on arriving at the most difficult number to ascertain, the
'forecast to complete'.

55. Dr Barnes is quite correct; the style is changing and the emphasis is to seek the co-operation of all departments working on a contract to get the right answer to this 'forecast to complete' exercise.

56. In answer to Mr Faulkner (paragraph 13), when a main contractor is appointed to handle a given contract he is responsible for the performance of all sub-contractors down to the smallest of them. This involves having expeditors who visit all companies as soon as an order is placed. They check that the sub-contractor has a programme, that this programme is sensible and that it will meet the overall requirements of the contract. Sometimes it is necessary for the main contractor to second a planner to a sub-contractor to help him establish a programme that will fit in with the requirements and the reporting procedures of the contract as a whole. Similarly it may be necessary to allocate an expeditor full time to a sub-contractor's works to ensure that he maintains resources and effort necessary to meet a tight programme.

57. As far as the involvement of top level management is concerned, in my experience programmes are regarded as a key document in the contracting business. They are reviewed with the same thoroughness as the price, technical content and conditions of contract when submitting a tender.

58. It is necessary to involve the heads of departments that are going to execute the contract and to get their whole-hearted commitment to the programme at the tender stage. This involvement must continue throughout the life of the contract and is generally achieved by having regular contract reviews of performance against the original programme.

59. In reply to Mr Berry, the 'S' curves included in the Paper are only typical. Each type of process plant will have its own shape of 'S' curve. During the tender stage, from the basis of the estimate and the tender programme, much detailed work must be carried out to arrive at the shape of the expenditure 'S' curve. The overall expenditure can be sub-divided into expenditure on head office services, expenditure on material supply and sub-contracts, and expenditure on civil, building and construction work. The performance of these three sections of the work can then be plotted against the target 'S' curves.

60. We have no experience on the idea of 'confidence bands' and so I cannot comment on this aspect.

61. By combining the expenditure 'S' curve and the receipt of money from clients (as laid down in the Terms of Payment clause) the cash flow can be determined. This will vary from contract to contract and similarly will vary between a positive and negative amount during the course of a contract.

62. In answer to Mr Postlethwaite, construction is the most difficult area as far as industrial relations are concerned and I believe that most contractors are now endeavouring to minimize the number of construction hours by prefabricating on to what are known as skid mounted units as much as they possibly can. This technique of necessity has had to be used on the North Sea oil platforms where large modules are prefabricated and then shipped and lifted on to the rig. Similar techniques, within the limitations of transport restrictions, are being adopted for process plants.

63. The other aspect of the elimination of industrial relations problems on construction sites is that facilities for the work-force have improved tremendously over the last few years and much more has to be spent on weather protection, scaffolding and safe working platforms.

Reference

1 BARNES N.M.L. Cost modelling — an integrated approach to planning and cost control. *Engineering and process economics 2 (1977).* Elsevier Scientific Publishing Company, Amsterdam, 1977, 45—51

7. Contractual issues – can the Contract help?

M.W. Abrahamson, LLB, FIArb*

Introduction

1. There is one test of the value of a written construction contract. Does it help to produce speedy and economic completion of the project, with as few abuses, misunderstandings and disputes as possible and the solutions of any that do occur? A contract that does not pass the test, i.e. that does not influence conduct to the benefit of the project, is not worth the paper it is written on, however legally clever it may be. A contract may be positively harmful if it sacrifices that influence to the inertia of the legal or construction industries or to ulterior motives.

2. It follows that the concentration in contract documents should be on intensely practical provisions to deal with the practical problems. In contrast to some unsuccessful modern legislation, the Old Testament and the commentaries on it incorporate this lesson. Their 613 rules for daily conduct are a striking example of practical and concrete regulation in preference to reliance on abstract principles. (For the standard Conditions of Contract as holy writ see paragraph 29.)

Invitation to tender

Passing of information

3. It is the terms of the invitation to tender, about information passing from the employer to the contractor and vice versa, that partly decide at the start of the relationship between contractor

---

*Lawyer

and employer whether either is buying a pig in a poke. The primary purpose of giving and getting information is not to prevent or facilitate claims, but to cause each party to know as much about the project as possible, so that all necessary plans for contingencies are made in advance. That is why the invitation to tender deserves more attention than it usually gets. (Lawyers particularly tend to ignore it, because traditionally it is said that the invitation to tender is not part of the contract.)

Site information

4. How much then should the Employer tell? Some say he should tell all and take responsibility for all he tells:

> 'The premise that tenderers have means of access or the time to carry out any meaningful additional below ground investigation during the tender period is false and should be abandoned . . . It is recommended that no disclaimer or warranty is necessary or desirable. Clauses 11 and 12 of the ICE Form should be fully accepted and there should no longer be any doubt that the tenderer is entitled to take into account the factual data and the Engineer's interpretation of the findings provided with the site investigation report.'[1]

Some say the Employer should tell, but should add the traditional Irish rider 'mind you I've said nothing', by excluding from the Contract any guarantee of information, and clause 12 of the ICE Conditions of Contract.[2] Others take a more complicated position, as in clause 11 of the FIDIC Conditions.[3]

5. The choice between these alternatives should not be made on legal grounds divorced from the realities. (For my own views on some of the considerations that are relevant see reference 4.) The choice should aim to enable the Employer to give to the contractor, and to enable the contractor within reasonable limits to rely on, the information which the Employer is in the best position to find out, and to cause the contractor to discover the information that he can best find out for himself, for example that needed for his special methods of construction. In that way the project will start with the maximum combined knowledge about the site, on which those involved may safely budget and price, and perhaps more important, plan the works in advance.

6. The most difficult question is whether interpretation of factual data is something which the contractor should be given or left to decide on himself. I realize that those who have read my remarks so far will have been touched by their innocence:

> 'It has been suggested that certain tenderers may, to some extent, have allowed the concentration of their skills to shift too far into the managerial functions of their organisations and so leave the technical and engineering matters to insufficiently experienced staff. This implication is related to a view also expressed that site investigation reports are not adequately studied and applied at the tender stage and only seriously brought into use in order to sustain claims for extra costs when things have gone wrong.'[1]

The answer is that the conditions of contract and the information given itself should be written specifically (and interpreted by engineers and arbitrators) to discourage unreasonable deduction and extrapolation from information given, and if the sort of contractor mentioned in that quotation is on the tender list (I do not know why such contractors should be on tender lists for large projects) then the Employer can best use the Contract to help himself, the public and reputable contractors, by providing factual data only without interpretation or even adopting the above alternative of refusing to guarantee information and deleting clause 12. The contractor cannot plead equity against the Employer or the Engineer, because 'He who comes to equity must come with clean hands' — an old maxim of equity (in the technical legal sense, propounded by the courts of Chancery).

7. Finally the effect of the local law on the clauses chosen must be taken into account. Continental codes, and, for example, Middle East codes based on them, may put terms into the Contract. In any case the interpretation of the Contract under code systems may differ greatly from the results that would be achieved by Common Law judges.

Information from the contractor

8. Apart from information about the contractor's intended programme and methods, the information that may most usefully be obtained from the contractor by the invitation to tender, is the

breakdown of the contractor's rates and prices. With modern conditions of contract by which rates for variations and adjustments of payment for risks may be based on analysis of the rates in the bill, there is every reason for the engineer to know before contract how the contractor has made up his rates.

9. Contractors traditionally put forward two objections to coming across with that information. With the restless exchange of personnel that is common between contractors, the objection based on confidentiality has little strength. It is noticeable that the other objection, that because estimating is more of an art than a science and bulk changes may be made in the contractor's prices etc. he does not know how his own rates are made up, strangely very often vanishes when the contractor himself is claiming extra payment by reference to the alleged tender breakdown of his original rates and prices.

Delay

10. The essential point about liquidated damages is that they are not intended to be paid. Their real purpose is to influence the conduct of the Contractor by making it worth his while to bring on extra resources, improve supervision, etc., to save time rather than pay the damages. That is why it is such a serious step, to be taken by the Engineer only with his Client's full understanding and agreement, to insert low liquidated damages. At least that is so in countries where liquidated damages will be enforced whether more or less than the Employer's losses, for this is one of the cases where the written contract may have different effects depending on the legal system governing the contract. German law, for example, treats the damages stated as the minimum damages recoverable.

11. Unfortunately in practice liquidated damages are not always a very successful incentive. Some extra practical controls can be provided by the contract (and no doubt research and investigation would produce more, such as sanctions, expedition and bonuses).

Sanctions

12. The duties to furnish programmes and update programmes,

for example, in most of the standard forms have no sanction attached. The only efficient sanction is withholding or deducting from a certificate, but such a right must be carefully drafted if it is to be valid.

Expedition

13. In the USA it is common to have clauses that give the Employer the right to step in and order materials etc. or otherwise expedite the work at the expense of the Contractor where he is behind programme. The extent to which that action is a practical proposition is limited, but the right may be useful.

Bonus

14. It may be that the current practice relies too much on sticks and not enough on carrots. Unfortunately use of a bonus involves some difficulties. An Employer may not be pleased with his advisers if he receives the works late but nevertheless has to pay a bonus because the Contractor is entitled to an extension of time. It is an unfortunate fact that the existence of a right to extension, although it may be justified in the abstract, may reduce the incentive of the Contractor to struggle hard to minimize delays.

15. The possibility has to be considered of applying a more tightly drawn extension of time clause than the clauses now traditionally used for relieving the Contractor against liability for liquidated damages. (It will still be necessary to warn the Employer that if he or the Engineer holds up the Contractor, even if the Contractor is not entitled to an extension of time, he will be entitled to claim the lost bonus from the Employer as damages for breach of contract or, for example, as under clause 7 of the ICE Conditions.) At least stringent notice requirements are justified. There is the point also that where the ICE 5th edition is used, the Baxter formula appears to apply to any bonus.

Role of the Engineer

16. I may have said too much so far about regulating the

Contractor by the Contract, and not enough about regulating the Engineer and Employer. In fact a good example of the need to make sure that what is said in the Contract has practical effect is that although construction contracts traditionally enshrine the role of an independent and impartial engineer to keep the peace between the parties, the construction industry probably now has a greater proportion of contractual disputes than any other. A Contract that enshrines wishful thinking at the expense of the Contractor, is not likely to promote co-operation.

17. If the Contract is to help the project and not hinder it, it must surely be honest about whether or not the Engineer is an independent professional or the alter ego of the Employer. The 3rd edition of the FIDIC Conditions clauses 2 (1) and 69 (1)(b) has taken a step in the direction of honesty. The ICE Conditions still promise without qualification that the Engineer will be fully independent — e.g. 'in the opinion of the Engineer' (clauses 39 (1)(a) and (c), and 60 (2)(a)) — as emphasized in reference 5. Apart from the possible misbehaviour of the Employer, there is the fear that the Contract is unrealistic because in any competition between self interest of the Engineer and impartiality, the former may win.

18. With a complex construction contract for a large project involving high potential for dispute, and with the Contractor granting credit to the Employer, there is everything to be said for having one or more professionals trusted by both sides interposed between the Contractor and the Employer to avoid or reduce friction. If the cynics are right and modern pressures do prevent the Engineer filling that role (which I cannot accept), then those embarking on major projects must find another solution. Steps that have been taken in that direction include in the South African Conditions[6] provision for conciliation procedure (with lawyers barred!) before arbitration, and in other conditions use of an independent board, not otherwise involved in design and construction, to decide disputes.

Financing and ownership of materials and components

19. Obviously a large part of the cost of a major project is paid to financiers. To reduce the Contractor's financing costs it is common to provide for payment for materials and components

when, or even before, they are delivered on Site. Many standard contracts also provide for vesting of materials and plant etc. in the Employer as soon as they come on to Site. That may be an important security for the employer in large projects, possibly allowing him to reduce the amount of any bond required. A burden is therefore placed on the Engineer to ensure that ownership passes to the Employer effectively in law, particularly that ownership has passed to the Employer before he pays for the materials and components.

20. The famous Romalpa decision[7] establishes that where the writ of the English courts runs a clause is effective in a Contract by which it is stated that the ownership of the goods is not to pass from the seller to the Contractor or sub-contractor buyer until, for example, the whole indebtedness of the buyer on any account to the seller has been cleared, or even vesting in the seller title to products into which the goods sold have been converted or in which they are incorporated. *Nemo dat quod non habet* — no-one can give what he does not have himself — is the principle that applies. Such clauses are particularly important in times of economic instability.

21. It appears from oldish decisions (for example reference 8) that the Employer is protected once the goods are actually incorporated into the structure, because they then become part of the freehold free from the previous ownership despite any reservation of title clause. According to those decisions the Employer is in no danger of having his building razed to the ground by unpaid suppliers to the Contractor. Unfortunately before incorporation the other statutory protection for buyers does not appear to be effective in construction contracts. The Sale of Goods Act 1893, section 25 (2), protects a subsequent buyer without knowledge of the clause to whom goods, subject to a reservation of title clause, are delivered. The section does not protect the Employer because materials etc, will be delivered to the Contractor on the Site of which he has possession, and not to the Employer.

22. Extra clauses are necessary in the Contract, and they can be drafted so as to give practical and workable protection against these problems. Clause 54 of the ICE 5th edition already provides elaborate precautions to protect the Employer where payment is made for goods that are held off Site. The clause is not drafted in

too practical a way. Certainly the mere statement of the precautions on paper is no protection to the Employer if they are not put into practice by the Engineer. In drafting payment clauses and dealing with claims the Employer and Engineer would also be wise to take into account the truth realized even by the courts that 'cash flow is the very life blood of the (Contractor's) enterprise'.[9] A Contractor bleeding to death is not likely to be co-operative or reasonable.

Disputes

23. It is becoming fashionable among lawyers to denigrate arbitration. Many say that it is no improvement on ordinary litigation (a truly serious criticism). It is not so often mentioned that it is lawyers who have distorted arbitration to make it as similar to ordinary litigation as possible, and are now condemning arbitration for not being better than the model they have wrongly chosen for it.

24. The Institution of Civil Engineers is considering improvements in the Arbitration Procedure (1973);[10] the cause is a good one. Although only a small percentage of disputes actually end in a completed arbitration, the mere existence of a relatively speedy, inexpensive and penetrating system of adjudication would do much to discourage abuses. Such a result would increase the percentage of the time and concentration spent by highly trained personnel engaged on large projects on actually planning and constructing the project.

25. Full independent expert investigation or at least fact finding by an independent expert at the time of the dispute have been suggested for gathering the facts at the right time.[1] It is also essential to ensure that the clarification of the dispute that takes place for the Engineer's decision is not wasted and that the process is not repeated in arbitration. Lawyers too need to be regulated. The industry suffers by the desire of a corps of professional advocates to advocate as much, and with as much room for manoeuvre and surprise, as possible. The present procedure is that for arbitration everything starts from scratch. The dispute is redefined in special pleadings. Written records are read out to the arbitrator and put back by him into permanent form, during

a time at which at most one lawyer and one witness are active but about six lawyers and many more witnesses and executives are present and being paid.

26. On the other hand continental written procedure is by no means perfect. What is needed is a careful blend of english, continental, american and canadian procedures; it is the notion that one system must be retained or adopted in whole that is preventing progress.

27. No large capital project should be planned without grappling with this problem in advance. The stage is being or indeed may have been reached where by the end of such projects the facts are so confused and the claims so many and complicated that it is not a practical process to resolve them by arbitration. Settlements that as a result depend on exhaustion of one of the parties or coercion of the weaker party, are hardly in the interests of justice.

Records

28. Record keeping creates a great dilemma on large projects. Twenty years ago the problem in advising on construction was to persuade the Contractor's and Engineer's supervisors to write enough; now the problem is to stop them writing too much. The immediate problem is that constant disputation in correspondence and at meetings sours the atmosphere on the job and reduces co-operation. But that is only symptomatic of the deeper problem of some imbalance towards law in the industry between the ingredients that are normally involved in a co-operative venture — legal rights, moral rights, goodwill and reputation, and the wish to bring a joint effort to a successful conclusion. The deeper problem will be solved only when those concerned appreciate their long term interests — '. . . a greater number of law suits does not necessarily mean a higher stage of development of society (industry), nor better welfare enjoyed by the people living (working) therein'.[11] The following are possible improvements relevant to the immediate problem.

(a) The Contract should spell out a clearcut and effective system of record keeping independent of disputes so that furnishing the agreed records arouses as little passion on either side as is possible.

(b) Claims and disputes should be dealt with quite separately
from the execution of the works, so that the atmosphere of
dispute does not contaminate the whole relationship of the
supervisors. Obviously that will not always be possible since
the disputes are about the construction of the works, but even
then the lawyers' practice of not allowing the utmost vigour in
contention to affect relationships outside the immediate arena
is most beneficial.

(c) There should be restraint and diplomacy of a high order,
and balance between short term gains and long term interest.

Changes to the standard forms

29. Some supporters of standard contract forms venerate them as
fixed holy writ. In reality no standard form can legislate adequately
for the great variety of works. Thus it is necessary for the special
features of every large project to be catalogued by the Engineer,
and a decision made as to whether any changes or additions to the
forms are necessary, and by the Contractor to see if any qualification to his tender is desirable. A large proportion of such specially
drafted clauses unfortunately produce the opposite result to that
intended. It is not for me to say that there would be an improvement if they were drafted by lawyers, but it is certain that any
change must be written with caution, adequate time, full consideration of all the practical implications and repercussions on
other clauses of the contract, and without the pomposity and
jargon that is the hallmark of lazy legal drafting.

Contracts and contacts

30. Many large projects are at the moment being carried out in
countries where there is some impression that Contracts are less
important than contacts, by whatever means the contacts may be
made, and where what law there is is subject to more or less
arbitrary change. Some contractors and consultants entering those
countries take the view that it is not politic or good for business
to insist on legal 'formalities'. I am of the opinion that this view is
right only if legal forms are necessary merely to please lawyers and

courts, and not also, as I believe, because they help the parties to genuine agreement without misunderstanding, and to a genuine understanding of what they are letting themselves in for. If that opinion is right then in situations where misunderstanding is particularly easy, because of the different backgrounds of the parties, and where misunderstanding can be particularly disastrous, it is most important to follow good contractual procedure (however diplomatically that may have to be done) when a deal is being made. Even a despot is likely to act less despotically in the face of a clear record of his prior agreement on the point in dispute.

31. I predict that the legal professions of many countries will in the not too distant future be grateful to those who, mesmerized by the number of noughts in the figures with which they are dealing, fail to take precautions for large projects that they would at home regard as elementary for even the smallest contracts.

Implemented research

32. Much of this Paper amounts to no more than a case for conducting, and above all implementing, intensely practical research and study if large capital projects are to start on the right contractual basis. Many have been saying this for the past decade and are likely to be still saying it for the next.

33. This Conference and the publications to which it has been possible to refer in this Paper show that some hopeful developments have taken place recently, but that there is much further work possible that would well justify the cost involved; the major problem is implementation of results.

References

1 CONSTRUCTION INDUSTRY RESEARCH AND INFORMATION ASSOCIATION. *Tunnelling – improved contract practices.* Construction Industry Research and Information Association, London, 1977, Apr., Technical Note 82

2 INSTITUTION OF CIVIL ENGINEERS, ASSOCIATION OF CONSULT-
ING ENGINEERS and FEDERATION OF CIVIL ENGINEERING
CONTRACTORS. *Conditions of Contract and Forms of Tender,
Agreement and Bond for use in connection with Works of Civil Eng-
ineering Construction*, 5th Edn. Institution of Civil Engineers,
Association of Consulting Engineers and Federation of Civil
Engineering Contractors, London, 1973

3 FÉDÉRATION INTERNATIONALE DES INGÉNIEURS-CONSEILS.
*Conditions of Contract (International) for Works of Civil Engineering
Construction*, 3rd Edn. Fédération Internationale des Ingénieurs-Conseils,
The Hague, 1977

4 ABRAHAMSON M.W. Contractual risks in tunnelling: how they
should be shared. *Tunnels & Tunnell.*, London, 1973, Vol. 5, No. 6,
587—598

5 ICE CONDITIONS OF CONTRACT STANDING JOINT COMMITTEE.
*Guidance Note 2A: Functions of the Engineer under the ICE Conditions
of Contract.* Institution of Civil Engineers, Association of Consulting
Engineers and Federation of Civil Engineering Contractors, London, 1977

6 SOUTH AFRICAN INSTITUTION OF CIVIL ENGINEERS, SOUTH
AFRICAN ASSOCIATION OF CONSULTING ENGINEERS and
SOUTH AFRICAN FEDERATION OF CIVIL ENGINEERING
CONTRACTORS. *General Conditions of Contract 1972 (incorporating
Forms of Tender, Agreement and Bond) for use in connection with
Works of Civil Engineering Construction*, 4th Edn. South African
Institution of Civil Engineers, South African Association of Consulting
Engineers and South African Federation of Civil Engineering Con-
tractors, 1972

7 Aluminium Industrie Vassen B.V. v Romalpa Aluminium Ltd. 1976,
1, W.L.R. 676

8 Reynolds v Ashby. 1904, App. Cas. 466

9 Lord Denning, quoted in Gilbert-Ash v Modern Engineering. 1973, 3,
All E.R. 195, 215

10 *Institution of Civil Engineers' Arbitration Procedure (1973).* Institution
of Civil Engineers, London, 1976

1 *Law in Japan.* 1974, Vol. 7, 337

8. Contractual issues – the contractor's view

D.F. Carmichael, MA(Cantab)*

Introduction

1. It is, or should be, a sobering thought that in the space of one, two or (if he is lucky) three months, a contractor has to put a price on a project which has perhaps taken several years to justify on the grounds of viability, which has taken a consulting engineer perhaps six or nine months to engineer and design and which the contractor himself may take three years to complete. In debating the management of large capital projects it is worth studying in detail some of the major problems which a contractor has to face before he is awarded a project to manage. For the purpose of this Paper we will assume that a contractor (being thoroughly honest with himself) has, in his opinion, adequate financial and technical resources to undertake the project in the event of his tender being accepted. He therefore decides, having gone through all the trauma (and expense) of prequalifying and being placed on the list of invited bidders, that he will tender for the project.

Pre-contract studies

2. During the preparation of the tender a contractor is in effect required to plan the project in detail to enable him to make his best assessment of the real or actual cost. To accomplish this, the contractor must send a team to visit the site of the works to obtain whatever information may be required to enable him to plan and price the contract. This team may comprise the contract manager and/or the project manager designate, the estimator, the planning engineer, perhaps a geologist or a soils engineer, and perhaps rep-

*Chief Civil Engineering Estimator, George Wimpey and Co. Limited

resentatives from one or more of the specialist sub-contract companies whose price may have a major bearing on the tender. In the event of the contract being undertaken by a joint venture or consortium, the number of people on the site visit can increase in direct proportion to the number of companies participating.

3. Every contractor has his own procedure for pre-contract studies, but for an overseas tender, whatever the procedure, he must attempt, in the space of days rather than weeks, to obtain the answer to every problem with which he may be faced in the execution of the project before any attempt can be made to start the preparation of the estimate. Only when he feels that he has obtained all the information which he requires, can the contractor's representatives return home to commence the detailed work which will be necessary to produce his tender.

Preparation of bids and tenders

4. The method of tender preparation is usually something which has evolved over the years, and is based on the contractor's past experience coupled with various check lists to ensure to the best of his ability that nothing is omitted. Bidding for a contract is essentially a team effort by all concerned — contract management, commercial, estimating, planning, surveying, purchasing, equipment-utilization, etc. In the case of large overseas contracts this team has to be widened to include the legal, insurance, finance and taxation departments. All these different activities, each with its own experts, have to be pulled together to produce the final answer. A mistake by any one of them, while perhaps not spelling immediate disaster, could have dire consequences for the Contractor in a project lasting for two or three years.

5. In planning a contract for tender preparation, a contractor will have to take into account numerous factors. If he tackles the job in a certain way he may need relatively more staff and more plant, but by doing so he may effect a saving in time compared with doing the job by another less costly method, which in itself will complete the contract by the required date. Which is more important to the Employer — time or cost? Another problem is plant utilization. A contractor may have a piece of equipment, the use of which would effect a significant saving in cost if perhaps

the design were modified. But will the consulting engineer be
generous enough to admit that the contractor can improve the
design, or will he be so prejudiced that he will not entertain
the consideration of any change? And then there is sub-contracting.
Is it more economical to sub-let or should the contractor do the
work himself? Perhaps the work in question is of a specialist
nature for which the contractor does not have the necessary
facilities and expertise or it may be politically wise to sub-let
part of the work. In this case will sub-letting the work affect the
final completion date, particularly if the sub-contract is on the
critical path?

6. Large capital projects always require the Contractor to provide
a considerable investment in construction equipment. In the UK
the problem is not too serious; if the Contractor already owns the
equipment he can include for its rental at predetermined rates, and
if he does not own it he can hire from outside, or he can purchase
new equipment if he anticipates further use on later contracts. On
overseas projects the picture changes. Unless he is operating from
an established depot abroad (in which case the situation is much
the same as it is in the UK) the Contractor must decide what
policy to adopt for his plant. Does he for example only charge a
minimal rental in the hope that he can use it later elsewhere? In
this case he must allow for shipping the plant out; but does he
also allow for shipping it back home or elsewhere (thereby perhaps
inflating his bid to his disadvantage) or does he take a risk and
hope that he can find further work in the country where he is
bidding. This of course is a risk which must be balanced against
the possibility of no further work, in which case the sale of the
equipment at a price below its depreciated value is a possibility.
These and many other matters have to be studied, considered,
debated and cost evaluated before the final cost price can be
achieved.

Responsibilities of management

7. In finalizing the price to be submitted decisions must be taken
on commercial, economic and perhaps even political grounds.
Management must consider what other competition there will be,
whether there will be any further work at the conclusion of the

contract, and indeed even the financial status of the Employer. Management must face up to the fact that some overseas countries are not as stable as others, and that by taking a contract there is an inherent political risk. The technical problems must be reviewed and appropriate contingency allowance made where this is required and, where fixed price contracts are requested, a decision has to be made as to the likely effects of inflation and currency fluctuation. The legal and contractual obligations of the conditions of Contract must be considered, and qualifications made where acceptable.

8. Finally there is the mark-up. Once this was only a question of an allowance for overheads and profit, but no longer. Large capital projects cost money to mobilize and initiate, and in times of high interest rates or financial stringency the cost of financing a project cannot be ignored. Terms of payment have to be assessed very carefully, and projected cash flows must be studied. The mark-up must include *inter alia* all the costs of financing and insurance (including bonding), and the former can be significant where payment terms are unfavourable to the Contractor.

9. The foregoing summary of tender preparation (albeit somewhat brief) brings to light several matters to which the contractor must address himself in hopeful anticipation of an award of contract, and which of course must be taken into account at the time of tender. Some of these are discussed briefly in the following sections.

Cash flow

10. The object of cash management is to ensure that cash inflows are available in the right amount and at the right time to meet obligatory cash outflows. However cash flow does not flow of its own accord — it can only do so as a direct consequence of management decisions, taken either consciously and positively or unconsciously by default. The major areas which could cause cash flow for good or ill are

(a) contract conditions relating to payment including down payments, interim payments, retentions, materials shipped, and materials on site

(b) operating decisions – the same range of decisions which contribute to profit

(c) capital expenditure decisions – the acquisition or disposal of plant, equipment or such other assets of a long lasting nature

(d) inventory decisions – changes in the amounts tied up in stocks and work in progress; increases create a negative cash flow

(e) client credit decisions – the length of time the client is permitted to take before he pays for sales invoiced or applied to him; an increase in client credit creates a negative cash flow

(f) supplier credit policies – the length of time taken before payment for materials, services and other items invoiced; an increase in supplier credit effectively creates a positive cash flow

(g) financial obligations – interest and dividend payments plus any contractual repayments of capital arising from past financial decisions.

The impact of the above groups will determine the net cash surplus or deficit at any point in time which leads to

(h) expense risk – other outside elements which can cause great impact include such items as inflation and currency rates. Since inflation is growth in value terms it is likely to have an adverse effect on the cash flow; similarly, currency rates determine the sterling value on conversion which will affect the cash flow

(i) investing decisions – the utilization of surplus funds (by the purchase of investment or similar) and the liberation of same when required

(j) financing decisions – borrowing on a short or long term basis from banks or other institutions.

Each contract will have its unique cash flow pattern which should have first been forecast at the tender preparation stage. The cash flow must be taken into account in the preparation of the estimate as this can affect the manner of pricing and/or the pricing of risk.

11. An additional consideration at the time of tender is the money at risk. This is particularly relevant for an overseas contract where cash value can be affected by both rates of exchange and currency regulations. The Contractor would not (without adequate safeguards) wish to be exposed to the draught of currency fluctuations by having net asset value position on a contract in currency. Similarly he

would not wish to finish the contract and find that surplus funds cannot be repatriated. The cash flow would therefore be better structured such that currency receipts matched currency payments, the balance of the receipts being received in sterling. Matching this requirement to the financing of the project might again lead to the best middle situation.

12. For the Contractor the control of the cash flow of the contract is a matter of month by month supervision. He will be concerned that the majority of preliminary expenditure is covered by the down-payment, and that the conversion of money spent to money received occurs as quickly as possible by ensuring that all work is measured and paid for promptly.

13. From the Employer's point of view the cash strategy of the various tenders must be closely inspected. The Employer will have established his source of finance for the project and pre-sumably evaluated its cost. It will be important therefore that the provision of money is at least level with the contractual requests for payment from the Contractor and also that provision is made for a reasonable sum to cover variations, escalation, and currency fluctuations. There is little point in either Employer or Contractor ignoring the cash flow of the other to the point of disrupting pro-gress on the project that has brought them together.

Cost escalation

14. In spite of world wide efforts to reduce the annual level of inflation, one of the Contractor's greatest problems is planning his operations to minimize the effects of the moving tide of cost escalation which will continue to flow for some time to come. However, the rate of such escalation is one of the great unknowns. Neither Employer nor Contractor wishes to take any more of the cost escalation risk than he has to, but at the present time both at home and abroad work is highly competitive, thus creating for the tendering contractor more onerous conditions than might otherwise be the case under more normal circumstances.

15. Cost escalation on civil engineering projects, particularly major projects of relatively long duration, is now probably the most important single factor in the preparation of a competitive fixed price tender — where it frequently rivals technical difficulty

and logistical problems. The more recent wide scale introduction, in the United Kingdom and elsewhere, of price adjustment formulae has both simplified and speeded agreement of cost escalation entitlements and more importantly for the Contractor the prompt payment of sums due to him. There has been some criticism of a blanket formula adjustment applying to individual work sections in a non-representative manner, but taken overall the introduction of such formulae has been a great improvement.

16. Many contracts overseas come out to tender on a fixed price basis which cannot under present conditions be in the best interest of the Employer, or in the longer term the construction industry as a whole. One gets for example a high set of prices, or a Contractor who cannot complete his obligations, or perhaps a government subsidized bid to produce 'hard' currency to the detriment of those contractors who quoted realistic prices. Any Contract which exposes the Contractor to risks which are considered to be unreasonable under normal practice cannot be in the best interests of the Employer.

17. Assuming that cost escalation is admissible on any overseas contract one may have to choose between a set of unreliable local indices, or the prospect of dealing with a traditional 'wage sheet and invoices' escalation calculation with an unsophisticated representative of a bureaucratic Employer. One further point is that with the rates of inflation experienced in recent years it is now unreasonable to expect that the Contractor should not be entitled to any overheads and profits on his recoveries when escalation can form as much as 50% of the original Contract Price. The cost escalation insurance in respect of fixed price overseas contracts provided by the Export Credits Guarantee Department (ECGD) and the private market have respectively proved inadequate and without sufficient capacity for large projects. They are not designed to suit the civil engineering contract where a major proportion of the work is carried out in the foreign territory. As with most serious monetary situations the solution as to who should take the risk on cost escalation will not be found easily.

The problems of bonding

18. The types of bond normally required are as follows:

(a) tender bond
(b) performance bond
(c) advance payment guarantee
(d) retention bond occasionally in lieu of cash retention (usually subject to negotiation).

The basic drawbacks of the need to provide bonds are that

(a) the cost of the bonds uplifts the price
(b) the bonds may well be counted against total financial facilities and so reduce the total that may be made available to the contracting company generally
(c) the bonds may be called through no fault of the Contractor and in meeting the call the Contractor may be seriously weakened financially, or even driven into liquidation
(d) there are circumstances when the Contractor bears a disproportionate amount of bond commitment compared to his share of the total contract
(e) the bond market in the UK is not sufficiently adequately developed.

The cost of bonds

19. The main sources of bonding are the banks. Employers overseas are increasingly insisting that bonds be provided through a national bank in the Employer's country. Since it is usually necessary for a UK bank to provide recourse for the foreign bank, two sets of bank charges result. The UK bank charges are usually much less than those of the foreign bank. If ECGD bond support is available the UK bank may insist that this be taken up, but a prerequisite of bond support is ECGD constructional works cover. When the contract is in a stable country, where constructional works cover is not otherwise considered necessary by the Contractor, this means a very substantial addition to the Contract Price.

Impact on financial facilities available

20. Banks look upon the face value of bonds as being contingent liabilities and therefore as part of the total financial facilities available. An important point to consider is the amount of bond-

ing in respect of contracts which have either been successfully completed or are in the tender stage. Thus, before tendering for a contract, it is important to ensure that any consequential bonding which might be required if the tender is successful, is available. The inability to provide a performance bond would probably result in the bid bond being called.

21. Even if a bond has expired it may not be cancelled by the bank until returned, in which case interest may continue to be charged. Difficulties are sometimes experienced in getting bonds returned. Undated bonds should, of course, be resisted. However, in the present buyers' market, any attempt by contractors to vary the terms of the bond unilaterally may result in the bid being rejected. Another circumstance where there may be a large bond commitment, which is not reflected in related turnover, is when contracts just finishing are overlapped by contracts just starting.

Potential liability of bonds

22. Bonds are of two basic types: 'conditional' where the onus is placed upon the Employer to prove default by the Contractor, and 'unconditional' which is payable on demand. In some cases conditional bonds may be effectively converted to 'on demand' bonds by the contract conditions, and also by the bank's desire to establish a clear cut call procedure which will preserve their international reputation for honouring their guarantees promptly and without demur. Malicious calling of on demand bonds is fortunately something which, so far, has not happened extensively, Libya being a notable exception. Even if an unfair call is not made the ability to do this gives the Employer a very strong negotiating position during the course of the contract to insist upon, for example, inequitable extensions or variations.

23. Even with ECGD bond support, the problem is not resolved. Having reimbursed the bank ECGD then look to the Contractor for payment. The Contractor may only avoid this or reclaim from ECGD if he can prove that the call was unfair. It may be difficult to achieve this, particularly if the remedy for settling disputes is taking an Employer to the local courts or grievance council.

24. If the bonds are expressed in a foreign currency, then additional exposure may arise. If sterling weakens against the foreign

currency, then the sterling amount of the bond increases. Also if a bond is called then it may be necessary to purchase the currency through the investment currency market rather than the official exchange market. For advance payment bonds, the bond amount should ideally reduce in line with the reducing amount of its outstanding advance payment. This however is not always permitted by the Employer.

Disproportionate bond commitment

25. Situations sometimes cannot be avoided where the Contractor bears a disproportionate amount of bonding compared to his share of the contract. There are occasions when it may be desirable or even necessary to take a partner or a sub-contractor selected for his technical expertise, but if he is not able to provide a bond then his share must be borne by the Contractor.

UK bond market

26. The bond market in the UK is currently presenting difficulties since the clearing banks are taking the view that because of the large sums now involved this is really not banking business. They are therefore looking to ECGD for support. If ECGD will not provide cover then it may be that a bond is only available, if at all, from a bank or insurance company on fairly onerous financial terms, for example depositing cash to part cover the contingent liability of the bond. A considerable difficulty is met in overseas contracts where financing is required, since ECGD support is only available for cash or near cash contracts.

27. The tendency for the future will probably be that the bonding institution, to protect itself in the event of a call that the Contractor cannot meet, will stipulate that large contracts are increasingly operated as financially autonomous entities. Consequently, any siphoning-off of cash to the parent company would need to be delayed to the later stages of the contract if a cash buffer was to be maintained in case of need.

Insurance of large contracts

28. The normal contractual risks to be covered are all risks

insurance, and insurance against damage to persons and property.

29. *All risks insurance.* This covers the Works, the Constructional Plant including temporary buildings during the period of construction and also in respect of work carried out during the maintenance period, and loss or damage incurred during that period due to a cause which occurred during the construction period. As the Contractor is normally required to cover replacement cost the insurance cover should be arranged to take inflation into account. Whilst the standard policy does not cover defective workmanship or materials it should cover other work which may be damaged as a consequence of such defects as this is a Contractor's liability.

30. *Insurance against damage to persons and property.* This covers claims brought by third parties and also claims which may be brought by the Employer or Engineer. This has particular relevance when a Certificate of Completion has been issued for part of the works, and loss or damage occurs to such part arising out of the execution of the works for which a Certificate of Completion has not been issued.

31. In addition, the following insurances are normally effected:

(a) motor insurance
(b) Employer's liability/workmen's compensation insurance
(c) marine insurance
(d) aviation insurance

How the Contractor arranges insurance

32. This may be effected by individual policies for each Contract and thus having a policy tailored to meet the specific requirements of the Contract. The Contractor may arrange his insurance policies to cover all contracts undertaken during a specific period of time, normally twelve months. This lightens the work load on the Contractor and ensures that he has automatic cover. It does mean that high risk contracts are under-priced whereas low risk contracts are over-priced so far as premiums are concerned. An adverse claims experience on one contract can affect all future premiums under a Blanket Policy. Also a serious breach of a policy condition on one contract may adversely affect the cover effected in respect of other contracts under the Blanket Policy.

33. Whilst a Contract normally requires a *minimum* amount of third party insurance the Contractor's liability is nevertheless

unlimited and he must assess his likely exposure at the Contract Site and surrounding areas, taking into account the value of any part of the Works for which a Certificate of Completion may be issued prior to a Certificate of Completion being issued for the whole of the Works.

34. Although the standard ICE[1] or FIDIC[2] Conditions are reasonably equitable, the Contractor must carefully examine the effects of any amendments to such Conditions. It must be remembered that not every liability is capable of being insured, and even certain liabilities which are capable of insurance may well prove uneconomic in terms of premium costs. Where the Employer is providing all risks and third party cover it is essential that the Contractor carefully examines the extent of such cover so that he may make it equivalent to his normal cover, either by pricing the additional uninsured risk or taking out insurance to supplement that provided by the Employer.

35. In many cases it is a legal requirement that insurances on overseas projects be effected with insurers operating in that particular country. The cover offered is often more restrictive in extent and the premiums are frequently more expensive than those of the international market. Where there is a shortfall between such local insurance and the liabilities imposed on the Contractor under the Contract, the Contractor should either provide for the shortfall in his price, or alternatively purchase additional insurance from the international market to cover such shortfall. Complications may ensue, as a result of the latter, from the settlement of claims for one loss by two insurers in different countries.

36. In the case of 'jumbo' projects the Contractor could find difficulty in placing cover for full value, particularly in the case of a project where water is present as in a marine or dam project. In such circumstances the Contractor will take out a policy for the maximum sum available against any single loss.

37. Where Employer-supplied free issue materials are to be the Contractor's responsibility the replacement value must be ascertained from the Employer and included in the insurance cover, together with the time of commencement of Contractor's liability. The taking into use of a part of the works by the Employer can have the effect of terminating the policy either in whole or insofar as that part is affected. Whilst the Employer normally accepts responsibility for claims arising out of his use of a part of the works

a claim could arise as a result of the Contractor being negligent where he was still working in that part. In that event he would not be insured and as the claim did not arise out of the Employer's use the Employer would not be responsible. In the event that the Employer is to take into use part of the works it is essential that the Contractor checks that his policy covers him for such eventuality.

References

1 INSTITUTION OF CIVIL ENGINEERS, ASSOCIATION OF CON-SULTING ENGINEERS and FEDERATION OF CIVIL ENGINEERING CONTRACTORS. *Conditions of Contract and forms of Tender, Agreement and Bond for use in connection with Works of Civil Engineering Construction*, 5th Edn. Institution of Civil Engineers, Association of Consulting Engineers and Federation of Civil Engineering Contractors, London, 1973

2 FÉDÉRATION INTERNATIONALE DES INGÉNIEURS-CONSEILS. *Conditions of Contract (International) for Works of Civil Engineering Construction*, 3rd Edn. Fédération Internationale des Ingénieurs-conseils, The Hague, 1977

Discussion on Papers 7 and 8

Mr Abrahamson

1. 'Law management' is bad if it dissipates goodwill and co-operation between the parties. The view that all problems can be cured by a return to contract forms of Victorian strictness is therefore an oversimplification. The contract must have regard to the record and character of those who have to be managed.

2. It is vital to the employer and to a good contractor that disruption is minimized and does not sour the whole contract. Harassment from late and exaggerated disruption claims is also a problem. Therefore, the contract should ruthlessly invalidate late claims.

3. Instead of trying to describe the notice of disruption and claims required in the usual convoluted language the contract should incorporate specimens of charts, notices and graphs which the contractor is required to complete during the works. Management tools will then be produced, capable of minimizing disruption by a co-operative plan of campaign tailored for the particular problem as soon as it appears, and should minimize conflicts of evidence about the disruption that does occur. Similarly energies should be directed to collecting, either by designated fact finders on each side or by an independent expert, 'real', firsthand evidence by photographs, tests and so on — rather than confusing and strident correspondence, by which each party concentrates more on trying to build a file than on building the works. The contractor's co-operation should be matched by a system of prompt payment for unavoidable cost.

Dr N.M.L. Barnes *(Martin Barnes & Partners)*

4. The standard conditions of contract (reference 2 of Paper 7) are still in the era where the contractor is assumed to be left alone during the construction of the work. On a civil engineering contract, particularly where it is part of a large project, it is unreasonable to think that the contractor can be left alone. Conditions of contract are needed which are more robust in keeping the project going and the contractor's incentives unimpaired when changes are initiated by the employer or necessitated by unforeseen circumstances.

5. To develop in this direction contractors could be required to declare more of the significant assumptions they have made about methods of construction than is now conventional. The new *Civil Engineering Standard Method of Measurement*[1] is a step in this direction.

6. Martin Barnes and Partners is associated with a project in the Middle East where the bill of quantities has been dispensed with. Instead a priced method statement is being used to provide a more realistic and uncontentious mechanism for dealing with the changes which occur.

7. The allocation of risk has been discussed in Papers 7 and 8. Conditions of contract appear to develop as a result of a struggle to pass the risks in the construction work to the other party. The contractor and the employer would each like the other to carry the consequences of the inherent uncertainties in civil engineering work. A recent example of this was where a major private client invited contractors to submit proposals for special conditions of contract to cope with especially difficult and tightly timed reconstruction. The client wanted the contractors to offer firmer than usual commitment to completion date, but instead more protection than usual from late information supply was requested.

8. Martin Barnes and Partners and the University of Manchester Institute of Science and Technology are collaborating on research, taking an overall view of construction risk allocation. The aim of the research is to establish criteria for allocation of risk based on the financial implications for client and contractor. If such criteria could be accepted, development of conditions of contract could become a more rational exercise, relying less upon tradition and the difficult concept of fairness.

Mr F.A. Fisher *(Rendel, Palmer & Tritton)*

9. Little has been mentioned about the role of the consulting engineer in the management of a project. Reference is made in Paper 5 to the engineer but this clearly means the contractor and not the consulting engineer. This may indicate the present tendency towards package contracts and the elimination of the consulting engineer in his traditional position. However, there are many examples where the traditional relationship of client, engineer and contractor still exists.

10. Most large projects inevitably involve a number of different contractors because of the variety of disciplines involved and the increasing amount of specialization. It is therefore preferable to appoint one major contractor to be responsible for the whole of the works and permit or instruct him to employ sub-contractors as necessary, and by doing so admit that the management of the whole project as regards co-ordination and programming will be borne by the main contractor; or, should the work be divided up into separate parts and stages, and separate contracts let for each part? There appears to be a majority of opinion in favour of the main contractor method but I am not convinced that it has any definite advantages.

11. The reason for adopting any particular method of management must first be established and assuming that the quality of work remains equal the principal criteria must be the final cost of the project to the employer.

12. It does not appear to me that individual contracts let separately should be any more expensive to the employer than the same works carried out as sub-contracts to the main contractor. The main contractor must include in his pricing an element for letting such sub-contracts and for the management and co-ordination of the project as a whole. If the consulting engineer is retained on a percentage basis the employer will, of course, pay more to the main contractor and a percentage of that to the engineer. If the engineer is employed on a time basis and the separate contract method of work is adopted, the engineer will have more work and therefore his fees will be higher. However, the general contract price should be lower because it will not have the main contractor's co-ordination and management element.

13. Interference of one contractor with another and additional or varied work will cause increases in the cost during progress of

the contract. Theoretically, by appointing one main contractor, responsibility for the additional work or default of a sub-contractor is carried by the main contractor and the employer is thereby protected. However, the main contractor may well add something to his price to cover such disruption and when faced with additional cost or delay on the part of a sub-contractor, he will try to recoup this in some way. Some main contractors have tried to pass on a sub-contractor's claim directly to the employer by the engineer. If the contract is properly drawn this can be firmly resisted, but if the main contractor simply substitutes his own name for that of the sub-contractor the claim must be considered like any other claim. Even if the sub-contractor encounters additional work and puts in a small claim, or does not claim at all, there is nothing to prevent the main contractor putting in a claim under the main contract, although there has been little or no additional cost to himself.

14. In the event of delay the main contractor is unlikely to submit tamely to liquidated damages, but will surely prepare a claim on his own account for delay, followed up by one for general delay and disruption, which is very difficult to adjudicate on.

15. If the separate contract method is adopted, claims by one contractor on account of another will still occur through for example, delay or interference, but this does not seem to make the position any worse than the main contractor method. It does put the management and co-ordination of the project squarely on the shoulders of the consulting engineer but there is no reason to object to this.

16. In the situation where nominated sub-contractors are used, there are recorded cases where a main contractor has been able to absolve himself from responsibility because a contractor has been nominated by the engineer. In this case, therefore, there is more reason for separate contracts.

17. Taking this to its logical conclusion, the use of provisional items and PC sums should be examined. These lead to problems; moveover the only reason for their inclusion must be that either the client has not made up his mind what he wants, or the engineer has not had time to prepare a detailed specification. They should by all means be avoided.

18. The above remarks have been made from the experience of completing two large and similar projects, one controlled by the

main contractor method and the other by separate contracts, from which it was concluded that in the employer's best interests the consulting engineer should call for separate tenders for distinct or specialized parts of the work and undertake the management and co-ordination of the whole project.

Sir William Harris

19. I understood that in recent years there had been a tendency towards management contracting: the main contractor acting as the manager. In this case, are the sub-contracts or direct contracts made between the individual contractors and the client legally, but managed by the main contractor?

Mr P.A. Thompson *(UMIST)*

20. Figures 1 and 2 illustrate the magnitude of the risks to be taken and the judgement demanded when compiling an estimate and preparing a tender for overseas work.

21. The cash flow curve for a large earth dam in Africa (Fig. 1) is typical of a job for which little material is purchased but high mobilization costs are involved for construction plant; thereafter all costs are seasonal and time-related. In this sort of contract there is great opportunity for negotiation on the amount of the mobilization fee which is frequently fixed from rather arbitrary rules. The increase in mobilization fee shown in this diagram would lead to a reduction of about 5% in tender price.

22. Some of the risks involved in a pipeline contract in the Middle East are compared in Fig. 2. A 30% change in the assumed inflation rate would increase the contractor's costs by 7%. If the contractor is 40% out in the residual value of his plant, the total cost of the job would increase by 10%. Risk contours such as the 30% probability line on this diagram are useful when assessing the total risk on a contract.

23. The estimates and risk analyses for the dam and pipeline were in fact produced for the Consulting Engineers, Howard Humphreys and Sons, on behalf of the funding authorities. If the client or his engineer is using such information to assist with

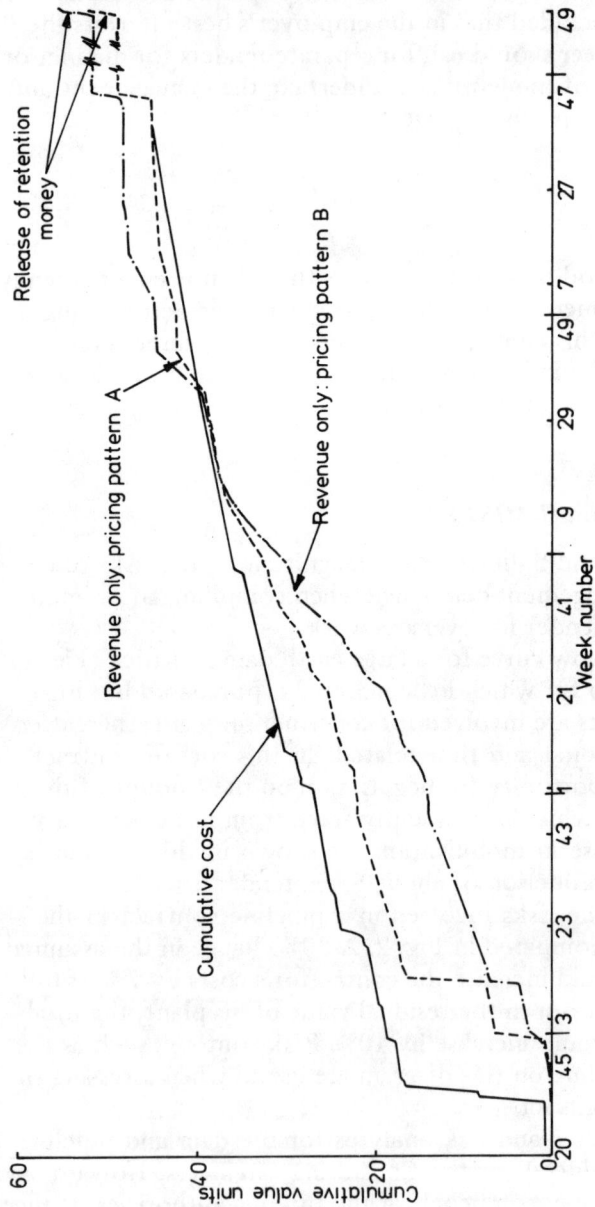

Fig. 1. Earth dam: cumulative cost and revenue; effect of mobilization payments

tender appraisal and subsequent contract management, there is little point in contractors hiding the build-up of their rates, and identity of purpose might be achieved.

24. Many issues were raised in Paper 7 which affect traditional contractual procedures. The engineer, in numerous cases, is unable to exercise his independence, although this situation is not necessarily reflected in the conditions of contract. More contracts are negotiated; the allocation of risk varies between contracts and there may be excellent political or logistical reasons for involving the client directly in the management of the contract. Knowledge of the success and the problems associated with each different approach to contractual arrangements is essential if British firms are to be successful in a highly competitive world market. More publicity is needed, and further research into contractual methods and issues. The CIRIA/UMIST study of target contracts[2] is a realistic and practical approach to such research.

Mr A.W. Shilston *(Consulting Engineer)*

25. A justifiable indictment of the professions, generally, is a reluctance to engage in constructive self-criticism with a view to achieving self-motivated improvement of its current practices. Paragraphs 23 — 27 of Paper 7 are salutary in that they question the role of the legal profession in the practical arena of the performance of construction contracts.

26. In the conduct of arbitration the specialist lawyer has a potentially creative role to play in helping the construction industry to manage its affairs. Under present conditions lawyers, with no alternative to offer but a parody of court litigation procedures, should not be allowed to dominate the procedural conduct of arbitration.

27. Former practising lawyers of the highest distinction, such as Lords Devlin and Shawcross, and even practising lawyers articulating through the activities and publications of 'JUSTICE', the British Section of the International Commission of Jurists, severely criticize English litigation practice. What is not required in the resolution of construction contract differences by arbitration is the unimaginative importation of discredited litigation procedural practices which often degenerate into an exercise in intellectual gamesmanship.

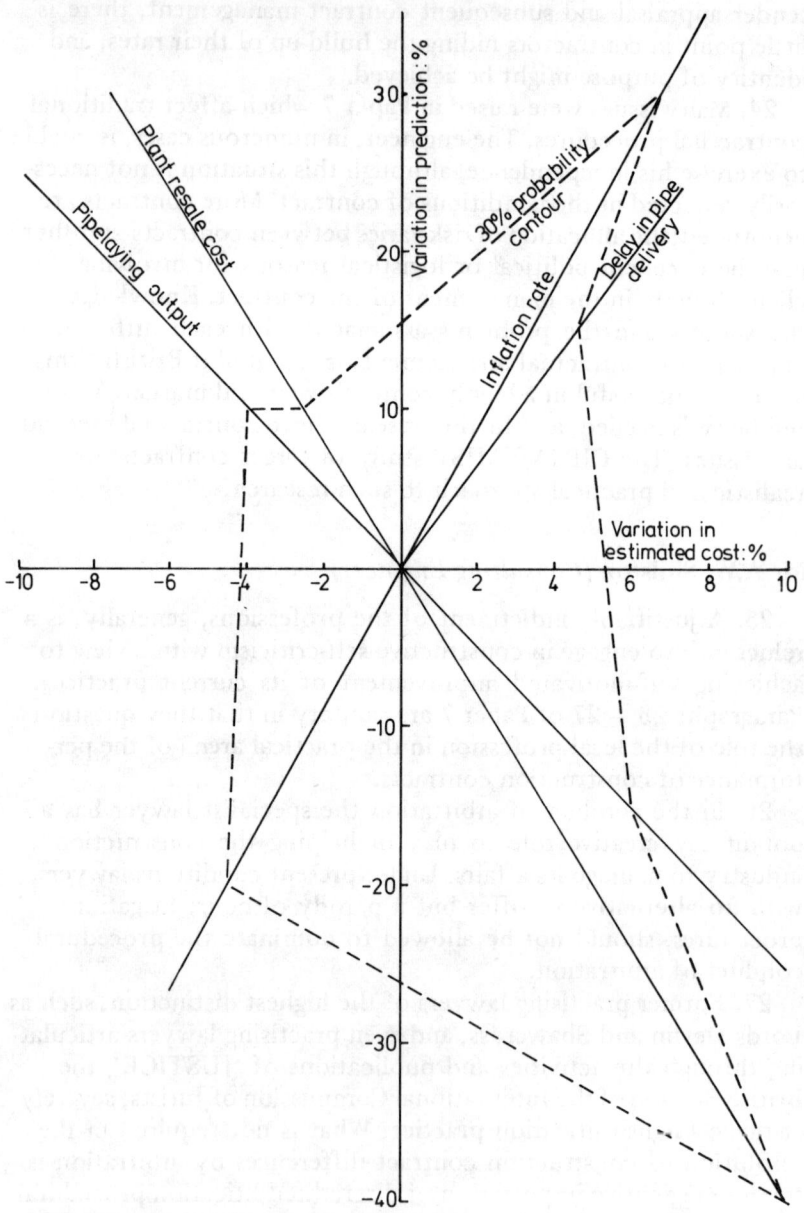

Fig. 2. Risk factors – pipeline contract

28. Can the Author elaborate on the ideal procedure mentioned in paragraph 26? Continental procedure epitomizes a system in which the arbitrator appears to adopt a positive inquisitorial attitude which is alien to the English procedure modelled on court practices. A more interrogatory attitude on the part of arbitrators ought to enable the course of the arbitration to be determined not by the rival advocates but by the arbitrator himself and result in expedition by the judicious exercise of technical insights.

29. In the USA, a highly developed procedural system has been evolved by the institutionalized and powerful American Arbitration Association. Presumably it is this aspect of American arbitration practice that the Author finds commendable. The conduct of litigation in the USA would hardly seem likely to provide examples of helpful practices to consider importing into an idealized system.

30. I am not aware of any special contribution that Canada could make in the development of commendable arbitration practices. What improvement to existing UK practices could be made from that quarter?

31. Paragraph 7 of Paper 8 and paragraph 10 of Paper 12 refer to the qualification to tender bids. Presumably to simplify the comparison of tenders submitted, project sponsors sometimes make it a mandatory requirement that clean bids only are to be submitted. Using the evidence of Bosphorus Bridge as a backcloth, is it in project sponsors' best interests to preclude qualifications to tenders by insisting on the submission of clean bids?

32. Would the Authors please indicate, in general terms, how widespread within an international spectrum is the prevailing requirement to submit clean bids? On the face of it a contract resulting from a concluded negotiation based on a dissection of the technical qualifications accompanying the original tender bid is more likely to reflect a balance of the mutual interests of both parties to the emergent contract.

33. What advice do the Authors offer to potential sponsors of large capital projects regarding the practice of barring the inclusion of qualifications with tender bids?

Mr T.D. Kershaw *(Consultant)*

34. In Paper 7, the objective to which the Author points would

be achieved more easily if the engineer drafting a Contract supplements his draft with informal notes stating the objective of his draft clauses, so that any legal advice and textual amendments are given to him with a clear mutual understanding of the desired end result.

35. The ICE Conditions of Contract and FIDIC Conditions of Contract (references 2 and 3 of Paper 7) have the same basic philosophy — 'a fair distribution of the risks inherent in works of civil engineering construction'. Where circumstances have forced a consultant to enter into an agreement with the employer which restricts his capacity to fulfil all the functions of the Engineer on which FIDIC is based, this fact should be made clear to the contractor before the Contract is signed.

36. The best designed Contract cannot make provisions for all possible eventualities in the life of a major construction contract. Should unforeseen or unforeseeable circumstances arise which would have a major effect on the contract and the sharing of risk originally foreseen, it is improper for either party to refuse to recognize the situation and hide behind the text of the Contract. In such circumstances, it is the Engineer's duty to assist both parties in finding an equitable solution to the problem by modifying the Contract or developing an 'extra-contractual' settlement. Facing up to this difficult but necessary task has, in my experience, been in the best interest of the Employer, been fair to the Contractor and minimized or even cancelled out delays in completion of the works.

Mr Abrahamson

37. I agree with the points made by Martin Barnes. Control and information go hand in hand — if the engineer does not know what the contractor intends to do he may disrupt the contractor by the way in which he exercises control, particularly control necessary in order to make changes to the project.

38. I can claim to have drawn attention five years ago to the need for research into the allocation of risk, and I am glad to hear that such research is taking place.

39. Similar investigation and research is essential before any conclusion can be drawn about the relative advantages of the

traditional main contractor/sub-contractor system compared to the letting of separate direct contracts. That investigation must come first and lawyers should be required to give effect to the results in a satisfactory legal framework, rather than lawyers and others looking unaided into their hearts and producing a straight jacket into which practice has to be squeezed with results as disastrous as the system of nominated sub-contracting.

40. All that Mr Shilston says about the shortcomings of procedure in arbitration is all too true. Lawyers have distorted arbitration to make it as like ordinary litigation as possible, and are now condemning arbitration for not being better than the model they have wrongly chosen for it.

41. Since the likelihood of having lawyers who are not arch conservatives on both sides in any dispute is minute, it is perhaps necessary for the ICE Arbitration Procedure (reference 10 of Paper 7) to go further than it does at present and impose a procedure tailor made for construction disputes, whether or not both parties agree to it.

42. Mr Shilston asked me to elaborate on what I regard as the ideal procedure, but that is an extremely large and difficult topic on which I hope in the relatively near future to publish some conclusions — for what they are worth.

43. As an elected honorary engineer I might be forgiven for following the example of other lawyers in yielding to the temptation to expatiate on practical matters on which such experience that I have as a lawyer is likely to be misleading. However lawyers do have great and genuine understanding of one practical point that has been mentioned — the importance of good mobilization payments — as you will appreciate if you have ever tried to mobilize a lawyer.

Mr Carmichael

44. Dr Barnes and Mr Thompson both raised the question of future changes in contract documentation, in respect of both the bills of quantities and the conditions of contract. While it is difficult to envisage any method of 'valuation' other than the bill of quantities (particularly having regard to changes in the scope of work during the contract), there is certainly a case for pricing a

method statement. Nevertheless, in view of the traditional conservatism of the Construction Industry, I believe it would take time to implement this method and both employer and contractor would have to be satisfied that the resulting method was equitable.

45. In those cases where the employer and/or the consulting engineer have attempted to modify either FIDIC or ICE Conditions of Contract (references 1 and 2 of Paper 8) the experience of my company (in tendering for new work) has been that the result has been nothing short of disastrous. We maintain the view that although these Conditions of Contract are not perfect, they have been evolved by experience over a period of years and represent the best solution possible whereby the interests of both parties are adequately protected. Any attempt to modify these, inevitably results in the whole concept being changed for the worse, with the opposite result.

46. The 'target' form of contract fulfils a very definite requirement, particularly under the circumstances which are described in the CIRIA Report No. 56.[2] However, this form of contract can usually only be 'sold' to a knowledgeable client who can understand the operation of such a contract and who can be convinced that he is ultimately getting value for money. Equally obviously, there are circumstances (particularly in the developing countries) where a target form of contract would be ideal, but as sometimes happens in the case of a relatively unsophisticated client, it is very difficult to convince him that a form of contract other than a lump sum can be to his ultimate advantage.

47. Mr Fisher posed the question of the role of the consulting engineer, particularly in the case of large projects involving a number of different disciplines. I do not believe that there is a 'text-book' answer as to whether one main contractor operating a number of specialized sub-contracts is preferable to several smaller packages, each co-ordinated by the consulting engineer. Each contract should be considered on its own merits (not forgetting the client's particular requirements) and the consulting engineer should be sufficiently good at his job to be able to give the client the right advice in any particular situation.

48. On the whole question of the management and co-ordination of large projects, I believe there is a danger of confusing the managing contractor with the main contractor operating with a number of sub-contracts. Like many other new innovations in

the Construction Industry, the role of the managing contractor is subject to different interpretations in different countries. To the best of my knowledge, the term originated in the USA where the managing contractor is really acting as agent for the employer and as such may not necessarily undertake directly any actual construction work. In this role he manages and co-ordinates the contract and places individual contracts with other contractors on behalf of the employer. Under these circumstances the role of the consulting engineer is primarily one of engineering and design, all of which may be completed prior to commencement of contract or under certain circumstances may proceed in parallel with construction.

49. The consulting engineer should be quite clear as to what his role in any particular situation is intended to be. In Paper 5 Mr Frame expressed the opinion that 'strong client guidance and control is essential'. While this may be true so far as the contractor is concerned, there is always the danger that a consulting engineer may not be able to fulfil his intended function when under pressure from a strong employer and this in turn may place the contractor in an untenable position. Mr Kershaw made the valid point that 'where circumstances have forced a consultant to enter into an agreement with the employer which restricts his capacity to fulfil all the functions of the Engineer, this fact should be made clear to the contractor before the Contract is signed'.

50. Mr Shilston has raised the question as to the advantage or disadvantage of unqualified bids. The incidence of the requirement to submit a clean bid occurs in my experience in only a relatively small proportion of the tenders which we prepare, and those mostly in the Middle East. The most obvious danger of submitting a qualified bid, where qualifications are not acceptable, is that the bid bond immediately becomes liable for calling. However, in a purely legalistic sense this can be circumvented by a form of words which are deemed to supersede the terms of the bid documents. The generally accepted practice of requiring contractors to quantify any qualifications is not unreasonable and must ultimately be in the best interests of both parties; this always pre-supposes that qualification is capable of evaluation. Nevertheless there are certain matters a contractor cannot quantify (e.g. potential loss on changes in the rate of exchange over a long contract period) and in view of the extremely competitive level of pricing over the

last few years, I contend that it is not equitable nor even possible for any contractor to carry this sort of risk in the level of profit at which he is currently able to obtain a contract.

References

1 INSTITUTION OF CIVIL ENGINEERS AND FEDERATION OF CIVIL ENGINEERING CONTRACTORS. *Civil Engineering Standard Method of Measurement.* Institution of Civil Engineers, London, 1976
2 CONSTRUCTION INDUSTRY RESEARCH AND INFORMATION ASSOCIATION. *Target and cost-reimbursable construction contracts — a study of their use and implications.* CIRIA, London, 1975, Dec., Report No. 56

9. Contributions from the behavioural sciences

Lisl Klein, PhD*

Setting the scene

1. There is a pattern in the way in which some sciences have developed: from intuitive application, through systematic investigation (science) to more knowledgeable application (technology). Men were using levers long before they investigated the principles on which they work. Having established the principles, they can now use levers more effectively. The development is similar in the matter of human relationships and behaviour. After all, for a long time people have been having relationships and co-operating in work, societies have evolved rules and customs and ways of having them kept, children have learned from the responses of their parents what pays and what does not, and armies have been inspired to commit suicide. Moreover, and this is a further stage in sophistication, people have speculated about these processes and institutions. Much has been written on law, government, education, the division of labour and, more recently, management and organization, on the basis of intuition and experience, and sometimes with great wisdom.

2. What is recent is the systematic, and therefore scientific, study of these things. It may not always yield better answers than the intuitive wisdom of the specially gifted, but then achievement in any applied science is to raise average standards of performance, not necessarily the standards of the outstanding individual. It also provides an essential means of testing the intuitive answers — too often for comfort they turn out to be wrong.

3. However, when one looks at the output from this kind of research in recent years, it would not be surprising if people in

---

*Tavistock Institute of Human Relations

industry found it bewildering, confusing, and even threatening
at times. Social scientists must sometimes seem to the engineer to
take a role which is either merely critical, or remote from real
problems in its academic concerns, or both. I would therefore like,
in this Paper, to structure some developments in the behavioural
sciences in a way that brings into relief the contribution they can
make in the industrial scene.

4. Much remains to be done in actually getting this contribution
made, or rather in getting it made in an integrated way. From the
very little that I know about civil engineering, I have the impression
that an important characteristic of the profession is the way it
makes use of a wide range of specialists, specialist knowledge and
specialist trades, linking these together in relevant temporary
alliances through the course of a project. If that is correct, then
civil engineers could have some important lessons to teach to
social scientists. For in the social sciences we too have some
important and relevant knowledge, experience and skills; but we
have only very patchily found ways of making these operational
and getting them into use.

The contributions

5. The area which has been most clearly made operational is the
one that is generally called *occupational psychology*. It is usually
concerned with *selection, training and vocational guidance,* and
is essentially analytical in character. A training programme depends
on analysis of the task, identifying the relevant skills and possibly
simulating the task. Selection also depends on analysis of the
characteristics of successful performance, and predicting for these
characteristics. Application in this area is the longest established,
usually through personnel departments. There are not many firms
which do not do something systematic about selection and training. On the other hand, there are hardly any firms which do anything systematic about the other areas.

6. The second group is about *ergonomics,* or the use of a knowledge of *human characteristics in the design of systems or equipment for human use.* Ergonomists have tended to limit themselves
to the more measurable, human characteristics which can be
experimented on, and therefore sometimes to the more trivial ones.

For this reason I have for a long time been keen on collaborative work between ergonomists and other kinds of social scientist. During my time as Social Sciences Adviser in Esso this led, for instance, to the redesign of Esso's refuelling facility at London Airport.[1] The situation was that after landing, aircraft parked at any one of about one hundred stands on the long-haul and short-haul 'aprons' of the airport. The turn-around time for most aircraft was about one hour, but one airline was already trying to reduce this to half an hour. In this time not only did passengers and fuel have to be loaded, but caterers, maintenance engineers, cleaners, etc., all needed to park near, and work on, the plane.

7. The job of controller in the fuelling station was to make sure that fuelling trucks reached the aircraft on time, and the worst thing that could happen was that he should be responsible for delaying an aircraft. For information about aircraft movements the controller had the arrival and departure schedules of the airlines which were Esso customers; minute-to-minute information about the actual approach and arrival of aircraft was received by two tickertape machines from the air traffic control centre; information about the specific fuelling needs of particular aircraft was supplied by electrowriter and by telephone links with the airlines; and information about the availability of drivers and trucks came from duty rosters and truck logs.

8. This job had been becoming increasingly difficult. The rate of traffic through the airport was increasing at about 15% a year. Traffic problems on the ground were increasing in proportion and one could 'lose' a truck in ground traffic for up to forty minutes. In addition, the company was very cost-conscious and kept to a minimum the controller's resources of trucks and drivers. Skill and motivation were not problems. In interviews with the controllers, it became clear that the importance of the job was clear to all and liked by all. They were very much identified with the success of the operation and had many ideas about its improvement. But all of them, in one way or another, complained of stress, fatigue, and inability to unwind. Inevitably some informal ways of coping were being found: 'When I know I'm going to get a delay, I phone my pal who's maintenance engineer for the airline. He'll pretend there's something the matter with the engine and start pulling it to bits.'

9. After detailed analysis of the system, redesign suggestions were proposed and considered in several cycles of discussions with all the

controllers and supervisors who were the potential users of the system. Then a simulated control room incorporating the new features was built in a laboratory, and simulation experiments were designed. The programmes were written from information supplied by management, supervisors, and controllers, whose experience of the job was essential for presenting an accurate picture of the system. Questionnaires and interviews tested opinions at different stages of the experiment.

10. The controllers and supervisors then came and worked in the simulated control room, testing and comparing four different methods of working under three different load conditions (the current load, the load predicted for the following year's peak period, and the anticipated load for five years ahead). The intention was that the people who would be operating the system could contribute their experience to its design. They would also be able to try out some ideas of the designers which were unfamiliar to them (and which they at first did not like) in safe conditions, to try a number of alternatives before deciding on a solution, and to have some idea of how long that solution would remain viable. The solution arrived at in this way was installed and, several years later, is still in use and well liked.

11. Simulation and the testing of alternatives are not, of course, new concepts in design; it is unusual, however, to find explicit and realistic attention paid to the people in the situation. Such methods can also be extended very much further to include users, or industrial relations and motivational aspects of work; for instance, the controllers' concern for scheduling rest-pauses and meal breaks for truck drivers emerged as an important criterion. It is quite possible to simulate some of the 'softer' aspects of social system functioning, such as the effect on a group which has interdependent roles when one member stays away or is slower than the others.

12. From the social sciences comes another kind of contribution concerning the *design of jobs and work-roles*. A central concept here is that of the *socio-technical system*. In other words, whenever a piece of equipment or a production process is designed, it is unavoidable that work-roles and interactions between people get designed at the same time, and that these then form an integral part of the system which is created. The research and thinking have mostly gone on in relation to roles in production, but the same principle applies to user roles, too. Historically, production

engineers have tended to regard equipment and task specifications as constituting 'the system' without taking account of the roles involved, while some behavioural scientists responded to this omission by emphasizing 'human relations' at the work-place, but in turn did not take account of the specific technology.

13. The essential interdependence between the social and technical factors was highlighted during some research on mechanization in coal-mining; briefly, the cycle of coal-getting consisted of three operations — cutting the coal, removing the coal which had been cut, and making safe the cavity which had been created by moving pit-props and other equipment. Under the old system, this cycle was carried out by small teams of miners who were multi-skilled (i.e. each member of the team could do any of the tasks) and self-regulating (i.e. what needed to be done was clear from the situation and they did it without the intervention of supervisors).

14. With a higher level of mechanization this cycle was extended over three shifts, each of which was only responsible for one part, consisted of between forty and fifty men, and was on a different piece rate. It was found that the workers no longer felt responsible for the completion of the whole job or had feelings of responsibility towards the workers on the other shifts. Each shift, optimizing conditions for itself, created and passed on poor conditions to the people who had to do the next task. The system now, instead of enabling them to co-operate with each other, created irresolvable conditions for conflict. In addition, all the controlling and co-ordinating activities now had to come from outside and above, because no-one working at the coal-face was in a position to know the whole story. The work suffered in various ways, absenteeism rose, and the expected increase in productivity did not materialize; the designers had treated the technical system as if it was self-contained, without awareness of its social components.

15. In a second series of studies the researchers came across a number of work groups which, with a similar technology, had organized themselves differently. With the co-operation of the pit manager, the men had worked out a system of work rotation both within and between shifts which overcame the problems mentioned, and which meant again that they could work autonomously — there was enough information in the work group for them not to need supervising from outside. It therefore became clear that a

single technical system may still allow for a choice of how to operate it, and some choices are better for the people in the system than others.[2]

16. The first explicit attempt to design and implement a socio-technical system was made in India at the Calico Mills at Ahmedabad.[3] The ideas were further developed in a series of job design projects in Norway, in the course of which a number of design criteria were formulated. These have to do with the growth, development, and functioning of the people in the system. In order to meet psychological needs, and thus generate motivation, work-roles need to be designed so that they

(a) make a perceptible contribution to the utility of the product
(b) utilize some skill which is valued by the community
(c) overlap with preceding and subsequent roles in the work-flow
(d) include feedback — some form of knowledge of results
(e) perform a task which can, in some sense, be perceived as a whole. If this is not possible, then interdependent roles need to be designed so that a group's task can be seen to be a whole
(f) lead to some kind of desirable future (e.g. through opportunities to learn, grow, move)
(g) include some contribution to, and control over, the setting of standards (e.g. of quality, quantity, methods, timing, etc.)
(h) contain some variety
(i) involve an appropriate length of job cycle, not the minimal.[4]

17. These criteria may be to some extent culture-based, and may not necessarily be applicable to all people at all times. Recently there has been substantial growth in developing new forms of work-organization. In Western Europe, cultural differences can already be detected between different European groups, which may be as important as those between Europe and other parts of the world. What matters is to find ways of incorporating the values of any particular society into the design of its technical systems. Design strategies are beginning to be evolved which make this possible, such as systematically testing alternatives and giving workers or other interested parties a role in evaluating production processes.[5]

18. At the same time it is important to recognize that such criteria are not entirely value-based and are functional, not 'merely humanitarian'. When a system is designed so that it is congruent with the way people function, the system itself functions better.

For example, in a small hospital the nurses were used to taking specimens that required pathological examination — blood, urine, etc. — from the wards to the pathology laboratory. The hospital was work-studied, and this custom was criticized as being an inefficient use of highly skilled nurses' time. An unskilled porter was introduced to carry out the task. As a result, relations between the wards and the pathology laboratory deteriorated sharply and the care of patients suffered. The scheduling needs of work in the wards and work in the pathology laboratory differed, owing to the nature of the work. However, when the staff of the two departments had met in the course of their normal duties (overlap of roles) they had each known enough of the other's situation to make, without explicit or even conscious effort, some adjustment in their own.

19. This brings us to a fourth kind of contribution, which concerns *appropriateness in organization.* For a long time people have been seeking general rules for how to organize. Ideas which have succeeded in particular situations (such as the hierarchical structure of an army in wartime) have been assumed to have a general validity, have been taught in management schools, and applied in situations which may differ in important ways from the original ones. There is a body of research in the social sciences which points to relationships between the structure created by an organization's environment on the one hand, and organizational behaviour on the other, and which therefore encourages managers to design their own organization according to the real facts of their particular environment rather than according to some abstract principle.

20. One such study, carried out at the South-East Essex Technical College, began as an attempt to assess the results achieved by the management training given at the College.[6] The researchers wanted to see whether businesses which were run according to a number of general principles taught were more successful than those which were not. To this end they took a hundred firms in the area and grouped them, according to a number of criteria, into three broad categories of success: average, below average, and above average. They then looked at the firms in some detail to see what characteristics of organization the successful ones had in common — searching for such things as the number of levels in the hierarchy, the size of the span of control, etc. However, no such patterns

were found. What the team did find was that if the firms were ranged according to the complexity of their technology — from unit production through small batch, large batch, and mass production to process production — then patterns of organization appeared which were common to each of these modes of production. Within each stage successful firms tended to be those which were nearest to the particular pattern common to the group, and unsuccessful firms tended to be those which deviated from the pattern, which suggests that technology requires its appropriate pattern of organization. This may sound obvious, but it cuts across a great deal of management teaching.

21. Another study with similar organizational implications was carried out in the University of Edinburgh.[7] This was a study of a number of firms in the electronics industry, which was at that time characterized by a rapidly changing technology and, in particular, rapidly changing markets. They were firms which had been used, during the second world war, to government contracts and now had to develop sales techniques and design techniques in relation to a completely different type of customer, who tended to express general needs only, and to need the help of the suppliers to translate these into specific requirements.

22. In these firms the researchers distinguished two kinds of management organization: mechanistic, with a clear-cut hierarchy, clearly defined roles and relationships, rights and duties, and this they deemed appropriate to organizations where things are not changing very fast, and organic, where jobs are not clearly defined, and lines of communication are fluid. This was found to be a more effective way of running things in the rapidly changing commercial and technical world of the electronics industry, where many concepts of good management, such as the division into specialized functions, the idea that a departmental manager should be self-reliant, and the idea of reducing skills at operator level, were more of a hindrance than a help.

23. It was found that the less successful firms were the ones which were slow to adapt their management structure in this way, partly because they did not recognize the need, but also because the mechanistic type of organization protects individuals and gives them security. The organic type may be felt as a threat, so that individuals tended to try to prevent the loosening up of traditional ways of operating. There are, therefore, observations in this study

about the internal politics and status problems of firms, as well as about the problems of integrating scientists into firms.

24. This study also highlights some of the problems of applying, or using, the results of research. In the climate of opinion prevalent in some circles in the 1950s and 1960s, the findings were thought to point to organic management organization as being 'good', and mechanistic organization as 'bad'. 'Organic, appropriate in certain kinds of environment and mechanistic, appropriate in other kinds of environment' was a less simple, and therefore less popular message. This is the kind of finding which cannot be translated straight into action. What it implies is diagnostic and adaptive still, rather than formulae for organization.

25. The same thing is true of the final type of contribution which concerns *process rather than content*. Any work group, design team, project team, organization, has in effect two tasks: to carry out its primary task, and at the same time to design, review and maintain the ways in which it does this. In practice this means that the way things are done, e.g. how an innovation is introduced, may be as important for the outcomes as the nature of what is being done. It means that the roles of specialists and the way they interact with their clients and with each other are important subjects for explicit consideration. It means that work groups and organizations may need to learn to review themselves, and may on occasion have to 'suspend business' to do so. One can be as careful and explicit in designing and reviewing these processes as one is in designing and maintaining products and structures.

Conclusion

26. It has become possible to take account of people's feelings and experiences as part of the data when looking at work situations; it has become possible to discover what appropriateness means in organization; it has become possible to create structures – tasks, organizations, systems – in ways that incorporate human and social, as well as economic and technical criteria; and it has become possible to pay attention to the process of development as well as to the outcome of development.

27. To social scientists, the value of all these contributions to the industrial scene seems self-evident. To others, they may present new

kinds of problems. There is a great deal which we still do not know, and which we must continue to explore, about how to make this kind of research and experience available and usable.[8] The blame for the discontinuity in application must be shared by both parties; on the one hand, many social scientists do not take enough trouble to convey their findings and methods in clear and operational terms; on the other hand, few things are more certain to evoke astonishment and horror among administrators and designers than the suggestion that in large capital projects — those, that is, that will have consequences for a long time — it is worth spending a small proportion of the costs (say one half of one per cent?) on R and D activities — research, experimentation, simulation, design — concerned with the people who will be involved in and affected by the project, their roles, relationships and experiences.

References

1 SHACKEL B. and KLEIN Lisl. Esso London Airport refuelling control centre — an ergonomics case study. *Appl. Ergon.*, 1976, Vol. 7, No. 1, Mar., 37—45

2 TRIST E.L. *et al. Organisational choice.* Tavistock, London, 1963

3 RICE A.K. *Productivity and social organization: The Ahmedabad experiment.* Tavistock, London, 1958

4 EMERY F.E. and THORSRUD E. *Form and content in industrial democracy.* Tavistock, London, 1969

5 WARNECKE H.J. *et al.* Neue Formen der Arbeitsstrukturierung im Produktionbereich. *Z. ind. Fertig.* 1975, Vol. 65, 665—670

6 WOODWARD Joan. *Industrial organisation: Theory and practice.* University Press, Oxford, 1965

7 BURNS T. and STALKER G. *The management of innovation.* Tavistock, London, 1961

8 KLEIN Lisl *.A social scientist in industry.* Gower Press, London, 1976

10. Towards better management of human resources

L.C. Kemp, MBIM, FFB[*]

Introduction

Manifestation of the problem

1. It has become the fashionable pastime of professional commentators to draw unfavourable comparisons between British
industrial performance and the much more superior productivity
and efficiency of our European, American and Japanese competitors. Unfortunately, these commentaries are based on reality
and the construction industry appears to be no exception to the
general rule. The recent study by the Economic Development
Committee for Engineering Construction[1] demonstrates disturb-
ingly poor performance on large engineering construction projects,
evidenced by long delays, widespread disruption, high manning
levels and low productivity. It has been argued by some that this
phenomenon is confined to engineering construction, possibly as
a result of the fact that large-scale plant suppliers make most of
their profit on manufacture rather than installation. Certainly
there is some evidence that civil engineering work on large sites is
not bedevilled by the same problems, but this is no cause for
complacency. There is plenty of scope for improvement on all
types of building and construction contract, and while the EDC
study described the problem 'writ large', its conclusions are no
less relevant to the industry as a whole.

Analysis of causes

2. The causes of this problem have been discussed and analysed

*Chairman, Construction Industry Training Board and Chairman, Corby
    Development Corporation

at length. They include errors and difficulties originating at the design and planning stage, the enormous problem of site management (directly related to the number of separate work-forces on site), and the characteristics of the operatives. The prime factor will, of course, vary from site to site, but a typical pattern is that a vicious circle is created which becomes very difficult to break out of. Often, the initial disturbance is an external influence which is outside the control of the Contractor (anything ranging from a Client's alteration of plan to bad weather). This disturbance factor may affect a vital stage in the sequence of work and cause delays out of all proportion to the original problem.

3. Individual labour forces which have been carefully scheduled to complete their sub-contract work in a particular sequence, instead become bunched together. At the same time, management may well decide to attempt to get back on schedule by offering incentives to selected groups and increasing manning levels, even to the point at which the labour force as a whole becomes impossible to manage. Grievances and disputes abound in this atmosphere, causing further disruption and delay. The attempt to recover the situation thus increases rather than solves the problem.

Objective of the Paper

4. The key to improvement lies first of all in reducing the incidence of initial disturbances which set up these damaging reaction chains. Often, these problems can be traced to the pre-construction stage and have very little to do with man management. It would be wishful thinking, however, to suppose that all such disturbances can be eliminated, and the next course of action, therefore, is to reduce their impact. It is here that the management of human resources can determine the success or failure of a project. The remainder of this Paper picks out the major areas in which effort should be concentrated if we are to avoid a repetition of past experience.

Manpower planning and programming

Project planning and overall management

5. One of the obvious features of large sites is the number of

separate stages which have to be completed, i.e. civil engineering, structural engineering, plant installation, construction of buildings, installation of services, instrumentation and control systems, and finishing. A typical large site can involve up to 100 separate work-forces. This immediately raises immense problems of overall planning and control. The observations of the EDC report were that management systems in the UK were just as varied as abroad, and planning techniques were no less sophisticated. But for some reason planning is less effective. Manning requirements are consistently underestimated and control of the flow of successive work-forces seems to break down more easily.

6. The history of the management of power station contracts in the UK provides an interesting illustration of the alternative approaches which can be adopted. Repeated difficulties in the 1960s led the CEGB to form a centralized Generation and Construction Division which adopted a system of appointing a handful of Contractors co-ordinated through a joint committee. This system, however, gave rise to its own problems. Using the absolute minimum of Contractors meant that some were responsible for work with which they were unfamiliar. Such a situation is just as likely to create extra difficulties as to secure improvements in site management. On the recent Littlebrook contract the number of Contractors forming the committee was ten, compared with five on the earlier Isle of Grain project. A further experiment on the Littlebrook contract has been to give the Client responsibility for site management, as well as design and construction on the main civil engineering contract.

Remedial action

7. Whatever the management system adopted, the overriding need is to plan in advance the timing of the arrival and departure of different groups of workers to ensure as far as possible that the labour force is fully utilized at all times. However, even the best laid plans can go awry. A significant difference between UK capital projects and similar foreign contracts was shown by the EDC study to be the ability to recover a situation following disruption of the timetable. It is at this point that a decision has to be taken on the relative advantages and disadvantages of increasing manning levels to get back on target, or at least prevent further

delays. There is some evidence that on UK contracts this strategy has been adopted without sufficient account being taken of the effects on efficiency and productivity of more intensive manning arrangements. At some critical point, and depending very much on the precise terms of the contract, it becomes cheaper to allow project time to over run.

8. In calculating the cost of alternative options it is necessary to have accurate information on productivity norms in relation to what is actually being achieved on site. The EDC report recommended the development of common standards of measurement, and this is certainly an area in which greater knowledge, and the dissemination of this knowledge, is required. Some sectors of the industry (e.g. electrical contracting) have already done a considerable amount of work in this field, and it is hoped that an extension of this practice will lead to a general improvement in industry's ability to develop more realistic and flexible manning policies.

Site capacity assessment

9. Accurate assessment of the gap between achievement and potential is important in determining whether or not to increase manning levels. Another consideration is the changing relationship between output and labour input. As more and more labour is added, problems of control will be such that extra labour will actually have an adverse effect on output. Before the project begins it is important for the management team to calculate the maximum capacity of the site, and to resist the temptation during the course of the contract to exceed this pre-determined limit.

Selection and employment policies

The construction labour market operating on large sites

10. A widely supported explanation of the lower productivity experienced on large capital projects is concerned with certain characteristics of the labour force which is attracted to this type

of work. It is claimed that a significant proportion of such workers are itinerant and rootless, have no company loyalty, are interested solely in maximizing their pay by whatever means possible, and take advantage of the anonymity of work on large sites to minimize their efforts. Some commentators even suggest that large sites offer perfect breeding grounds for political activists. A less extreme interpretation of the behaviour of site workers on large projects relates apparent sub-standard effort to low morale associated with the repetitive nature of much of the work, and the risk of discontinuity of employment arising from the hire and fire system. The knowledge that finishing the job means loss of work is not exactly conducive to enthusiastic activity.

11. The solution to this problem lies largely outside the control of the individual contractor, who has to accept the way the construction labour market operates as a fact of business life. The EDC report commented on the greater continuity of employment of workers in Europe and the USA but pointed out that the conditions contributing to this situation could not necessarily easily be imported into the UK. The European system benefits to some extent from geographical concentration of construction work, permitting the re-location of work-forces as demand varies from project to project. In the USA the unions in many cases act as hiring agents for contractors. Although labour is employed on a temporary basis, the union is under an obligation to find work for its members.

Decasualization

12. One step which it has been suggested that the industry as a whole should take is to set up a register of operatives working on large sites. The question of decasualization of labour in construction on the narrower definition (i.e. building and civil engineering) is a very topical issue, and could provide useful guidelines for action in the engineering construction industry. Although legislative attempts have not so far come to anything, the question has certainly not been dropped. Instead the present Government has set up the Construction Industry Manpower Board, under the chairmanship of Sir William Harris, to draw up proposals for stabilizing employment within the industry, if necessary with the aid of statutory powers.

13. Such a scheme would probably involve at least some of the following elements: attachment to the industry rather than the individual firm, registration of operatives, registration of employers who would accept an obligation to hire only registered labour, fall-back pay for unemployed operatives, and a redundancy payments scheme. Progress appears at the moment to be slow because of differences of opinion within the industry over how such a scheme might operate. While it is bound to impose significant financial costs, the effectiveness of this type of measure cannot be guaranteed.

14. An alternative approach is to bring pressure to bear on Government to stabilize the industry's work-load, thereby encouraging firms to maintain larger nuclei of permanently employed workers. Another important measure which could help to reduce the disadvantages of a casual labour market would be a properly designed pension scheme.

15. Finally, one of the most damaging aspects of the problem in terms of morale and labour relations concerns the use of 'lump' labour. Whatever happens to the decasualization proposals this feature of the labour market needs to be continuously monitored and controlled.

Skill standards, training and certification

16. One of the promising side-effects of the trend towards registration is its link with certification — implying a greater emphasis on skill standards, training and practical experience. This may well prove to have an important impact on the general level of productivity. An interesting discovery of the EDC's study was the lower proportion of skilled labour on UK sites, and the higher proportion of green labour. Basic training in construction skills has come a long way in the last decade, with an increased provision of off-the-job facilities and the design of training material. Moreover, many engineering construction skills can be acquired in a relatively short space of time, and courses can be mounted at reasonably short notice. With adequate planning, there is no reason why green labour should be used on construction sites, and if registration linked to certification can prevent this happening, this is clearly a step in the right direction. Schemes for scaffolders

and tunnel miners are already in the pipeline, and these will provide useful pilot studies for a more general application.

17. Although it is claimed by some that certification is bound to meet with worker resistance, there is also plenty of evidence that operatives welcome the opportunity to get their skills formally recognized. A recent scheme involving the grading of plant operatives introduced by a major civil engineering contractor apparently met with very little opposition. The electrical contracting industry has, of course, been operating a voluntary registration system based on grading for some time.

Supervision and site management

Selection and training of first line supervisors

18. A major determinant of the efficiency of the work-force on site is the effectiveness of site management policy. Considering first the problems of first line supervision, the EDC study suggested that numerical ratios were not as important as quality. It must be remembered that the supervisor is often at the forefront of relationships with other sub-contractors and labour forces as well as dealing with his own men. This aspect of the job may well be the cause of high turnover of large sites, as they succumb to the pressures of complex site relations. The selection of supervisors therefore requires careful consideration.

19. All too often the temptation is to appoint the best of the skilled workers to supervisory positions. There is no hard and fast correlation between technical skill and management potential, and the promotion of the most competent craftsman may in some circumstances not only deplete the productive potential of the workers on the tools, but may also add to management's problems if he identifies more with the work-force than with the management team.

20. Ideally, a common system of supervision should be established in the pre-construction planning stage. Irrespective of the experience of individual supervisors, it is probably to the advantage of all contractors and sub-contractors on site to participate in the setting up and running of a training course for supervisors geared to the specific project. This should cover not only the technical aspects

of the job, but also aim to instil an understanding of the commercial implications of site decisions.

Site management

21. Although problems of turnover amongst managers may not
be as great as for supervisors, there is evidence that the pressures
on long contracts are such that some managers have to give up
before completion through sheer exhaustion. The site manager is
generally the initial point of contact for industrial relations
questions and needs to have a strong grasp of the principles of
man management, industrial relations procedure, safety and employment law. In a rapidly changing environment, this implies
a continuing training need.

22. It has been suggested that a serious weakness in site management on large projects involving the installation of plant has been
the manufacturer's withdrawal from responsibility for the installation stage. There has been a tendency for directly-employed
construction work-forces to be replaced by small sub-contractors,
creating a multiplicity of work-forces and a management vacuum,
with missing links in the chain of accountability. If the quality
and efficiency of site management is partly determined by the
system of contract management, then the key to improvement
may be to insist on a contract package in which the manufacturer
undertakes to design, supply and install. Alternatively, the answer
may lie in construction contracts which are managed through a
co-ordinating committee controlled either by the client or the
main contractor. It is clear that, whatever system is adopted, the
location of ultimate responsibility and the chain of command
from contract management through site management to the
supervisor must be defined unambiguously.

Industrial relations

Introduction

23. Although project planning and general manpower policy are
important facets of human resource management, the single key
factor identified by the industry as a whole, and confirmed by

recent research, is the organization and conduct of industrial relations. It is difficult, however, to pin-point precise causes for the difficulties which are frequently encountered on large sites. Some commentators see the problem in terms of the organization of labour, citing inter-union rivalries and demarcation disputes, unreasonable claims for bonuses and severance pay, as evidence that workers are largely to blame for poor performance. At the other extreme, it is argued that management incompetence is the source of the problem, exemplified by an archaic approach to industrial relations which fails to recognize that workers are more than just 'factors of production'.

24. A third view is that the relationship between workers and managers is hampered by the fact that this is not always the point at which decisions can be taken. Management's hands may be tied by the client. These analyses are clearly too simplistic, and a solution can only be found by all parties taking a pragmatic and realistic view of the way site relations are operating and the scope for improvement.

National versus site agreements

25. Much of the discussion about industrial relations on large sites has centred on the relative merits of national as opposed to site agreements. In the UK different systems have been adopted for power station contracts, and oil and chemical plant installation contracts. On power stations, the practice has been to operate on the basis of national agreements, but supplemented by site arrangements covering certain specific aspects such as bonus schemes. This provides a useful source of flexibility, but often creates a climate of resentment between different groups of workers. The practice adopted in the plant installation sector has involved drawing up a uniform system of pay and conditions relating to an individual site. This avoids the problem of competitive claims between workforces; but success rather depends on how well-insulated the site is from the local labour market.

26. Often there is a combination of agreements in force. Even where national agreements are used as a basis, a typical large site will have three such agreements operating simultaneously: one each for civil engineering, electrical contracting and mechanical construction engineering. This has highlighted quite strong differ-

ences between labour relations in the separate industries which operate in this field. Contrasts have been drawn between the civil engineering operatives and the Mechanical and Electrical Engineering Construction Industry workers, with the former allegedly earning up to £2000 a year more on the same site. Superior conditions and better labour relations in the civil engineering sector have been attributed to strong organization, sometimes with 100% union membership. (On the CEGB Littlebrook contract, for instance, the entire civil engineering work-force, including managers, were members of appropriate unions under a post-entry closed shop agreement.)

27. One recommendation of NEDO's earlier report on large sites[2] was the institution of a *single* national agreement for large sites. This solution has a certain amount of appeal, but would nevertheless create difficulties of enforcement at local level. The fundamental determinant of success seems to be not so much the type of agreement in force, as its comprehensiveness in terms of coverage of the various aspects of site relations. It is claimed that in the US where a similar multi-union system prevails, labour relations have been assisted by advanced planning and comprehensive collective agreements.

28. An important feature of such an approach is the harmonization of individual contractors' conduct of industrial relations, which can be achieved in alternative ways. The client may appoint a resident industrial relations officer to be responsible for liaison between contractors and to act as a focus for management-union contact. Another method is for the main contractor to set up fully comprehensive procedures. An interesting model is provided by the National Exhibition Centre contract,[3] in which the main contractor (Douglas) introduced the concept of a site procedural agreement at the tendering stage.

29. Once the contract had been awarded, but before the work had begun, the contractor appointed an industrial relations consultant and held meetings with the appropriate national union officials from the National Joint Council for the Building Industry and the Civil Engineering Construction Conciliation Board to discuss a draft. An agreement was reached covering the appointment of shop stewards, arrangements for joint meetings, general conditions of employment, amenities and grievance procedures. The agreement also provided for a joint effort to ban the use of lump labour, with

the co-operation of labour-only sub-contractors. The initiative was successful because the nature of the agreement created helpful attitudes on both sides. The content of the agreement, as well as the manner in which it was established, was conducive to good industrial relations.

Payment systems, bonus and incentive schemes

30. On the question of the relative importance of basic earnings and bonus payments, the obvious preoccupation on large projects should be with equity between individual work-forces. Disputes arising from competitive or 'leap-frogging' claims can be reduced by sticking to a basic pay agreement which aims to maintain a steady relationship between the earnings of different groups, and to avoid wild fluctuations in pay from one week to the next (for example, by continuing to pay plus rates even when men are moved temporarily to lower grade work).

31. Bonus schemes are extremely difficult to fit into a pay struct-ure such as this unless they can apply to the site as a whole rather than to individual groups of workers. Civil engineering contractors have often been accused of 'buying off trouble' by agreeing too readily to bonus schemes for their own workers without sufficient thought for their effect on other contractors' work-forces.

32. The situation is helped if it is obvious to all concerned that a bonus scheme bears some relationship to productivity. This may reduce the tendency for copy-cat claims. It is important, therefore, to select the type of work which is suitable (such as repetitive installation work) and to set agreed targets which can be easily measured. Another approach is to focus on the quality rather than the quantity of effort, for example by grading workers on the basis of tests for specific operations or machines. Bovis have used a scheme of skill-related bonus, replacing the normal guaranteed bonus, with some success. The whole area of bonus payments is undoubtedly fraught with difficulties from the point of view of both management and workers, and the best strategy, as a general rule, is to maximize the proportion of basic earnings in the total pay packet. Nevertheless, bonus and incentive schemes will continue to operate and can, if used carefully, provide a useful source of flexibility.

Amenities

33. Another important element of a site agreement concerns the provision by the client or contractor of basic amenities, such as canteens, toilet facilities, lockers, drying rooms and first aid posts. An interesting fact emerging from the EDC reports was that site facilities tended to be better in the UK than abroad. Disputes over these 'hygiene' factors occur quite frequently, however, and the important point would seem to be the extent of their provision in relation to expectations based on experience on other sites.

34. There may also be difficulties if there is an unplanned increase in the work-force, causing facilities per man to deteriorate. To guard against grievances and disputes arising from this source it is wise to set out in the original agreement the level of facilities to be provided in terms of the number on site. Often, of course, grievances over amenities are a surrogate for a general malaise or low morale resulting from one or more other factors. They should therefore be investigated even if they appear to be trivial or spurious.

Procedures for disputes and grievances

35. Disputes and grievances are not going to disappear simply as a result of a more systematic and thoughtful approach to industrial relations. The final element of the site agreement must cover in detail arrangements for dealing with areas of disagreement. The agreement covering the National Exhibition Centre contract provided for the appointment of a shop steward from each trade and a convenor of all shop stewards who lived on the site and who was permanently available to members of the work-force. Similarly, the contract manager was available to the convenor of the shop stewards at all times.

36. This question of immediate access is a very important factor in resolving difficulties at an early stage before attitudes harden. Where a problem cannot be solved at this level, the agreement should define a clear sequence of events, including the specification of the time allowed for each stage, for the processing of a dispute up to the level of national officers, and independent conciliation or arbitration where necessary. There is no excuse for disputes to be complicated by interminable wrangling over the question of

procedure. An enormous amount of productive time can be saved by the careful drafting of a procedural agreement designed for a specific site.

Conclusion

37. This Paper has attempted to highlight and discuss some of the more important aspects of the management of human resources on large capital projects. It is difficult to summarize either the problem or the solution, because the circumstances and crises which confront managers and workers on site are as various and complex as the project itself. There are a number of techniques and guidelines which can be drawn on, but it would be foolish to imagine that these alone can provide the answer. They are simply a framework within which both sides have to operate. The scale of the task is such that management cannot be expected to maintain a constant vigil on every operation. In this situation, the principal objective of management should be to encourage the development of an environment in which the work-force is motivated to control and manage itself.

References

1 ECONOMIC DEVELOPMENT COMMITTEE FOR ENGINEERING CONSTRUCTION. *Comparative construction performance.* HMSO, London, 1977
2 NATIONAL ECONOMIC DEVELOPMENT OFFICE. *Report on large sites.* HMSO, London, 1970
3 THE BUILDER LTD. In place of strife. *Building*, London, 15 Apr., 1977

Discussion on Papers 9 and 10

Mr D.B. Parkes *(Managing Director, Tunnelling Division, Fairclough Civil Engineering Limited)*

1. The results of insensitive application of mechanization in industrial situations was brought out well in Paper 9. Civil engineering does not generally suffer from these effects because of the use of the gang as a unit, and this particularly applies to tunnelling.

2. There is, however, a marked change in behaviour when the size of a gang exceeds a critical value. What are the Author's views on the application of the group dynamics theory to the cause of efficiency. Small groups working from outside operate against productivity and good management, at present. Cannot proper use of available knowledge be made to work to our advantage?

3. The dangers of flooding sites with labour was mentioned in Paper 10. The most crowded site that I have seen was a metro station in Brussels, where apparently only the desired effects were achieved; but perhaps this was a short term situation. Even concentration of work-force is necessary, since areas of overconcentration may produce the same bad effects as the whole site being overloaded.

4. I cannot accept the enjoinder to avoid certain areas because of bad industrial relations. Civil engineers are required to construct a static facility in a specified area. Perhaps clients would be well advised to take note of the industrial relations performance of their contractor, possibly with some emphasis on the area in question.

5. With reference to the funding of international projects from banking sources, how far has consideration been given to the application of such funds to produce a more steady work-load in

this country? A sound base load is essential to any training scheme for operatives and engineers. It still appears necessary to remind some circles that an engineer's training is not complete when he attains his degree, and that proper home training in technical and managerial practice is an essential basis for overseas work. A way has to be found to promote a steadier release of work that does not bring about problems with training, management, and personnel looking to the consequences of the end of the job.

6. Safety is one of the major factors in the management of any contract, and consideration must be given to its promotion, along with industrial relations, right from the design stage. The UK will have legislation on this subject in the Autumn of 1978. Will this be an opportunity for true management-operative co-operation, or for exploitation by political activists and other disruptives?

Mr J. Bulman *(Director, General Site Services Ltd)*

7. In Jedda a few months ago I was told: 'British labour is lazy, difficult to control, fussy about its living conditions and too expensive'. Nevertheless, it was used in the end, and not unsuccessfully. But the image of British labour in this country and overseas is bad. It is not the exportable asset that it could be, and the image of the industry is damaged generally — not just its labour but also its management. We are all in this problem together.

8. We all have prejudices, no doubt, about solutions, and mine has to do with foremen. Paper 10 stated that foremen should be trained, but it also stated that the site convenor should have an open door to the project manager on every site. This quick channel of protest from the bottom to the top may be necessary and admirable, but it has always cut out the foreman. For many years the man on the site with a problem has been inclined to go through the protest system, and not through the system of management. How then can the management convey to the man on site its information, instructions, direction, motivation and inspiration? Unless the foreman is competent to convey it, nobody else will. Therefore, foremen must be selected, trained, encouraged and authorized to play their full part.

9. Contractors do not seem to tackle the foreman's role seriously. But, the man who pays the piper can call the tune. If clients and

consultants look at the way their contractors are set up to do the job of selecting, training, looking after and using their foremen, they can themselves improve the performance of the industry. It is not sufficient to look at the lowest price; that will not necessarily get the lowest overall cost. One must look at the efficiency of the work itself, and that depends on the efficiency of the foreman.

Mr D.S. Lawrenson *(Edmund Nuttall Ltd)*

10. The contractor's project manager — or the agent — actually carries out the work. The forces ranged against the agent these days are enormous and arise particularly from the welter of legislation, resulting in problems from the unbridled power of shop stewards and the inability of many union officials to control their activities. Many people do not really understand what industrial relations means these days and how the art is practised. The agent has to deal with both the other project manager, who blames him for the programme running late, and with the shop stewards and their different points of view.

11. Agreements such as bonus agreements and procedural agreements seem to be made for breaking. Do the Authors think the time has come for major agreements in this country to be made legally binding, like they are in other countries.

12. Are the days of the shop steward numbered? Would it not be better, instead of having shop stewards, to have more and better trained union officials, who could receive advice at the branch meetings, and then come forward from time to time with the various employers to discuss these matters?

13. Can appropriate clauses be written into contracts to protect contractors from the unreasonable industrial action which occurs these days? Can some of the current legislation be reversed?

Mr R.J. Bridle *(Department of Transport)*

14. The papers deal largely with relationships between management and work-force. While these relationships are of fundamental importance, other intergroup conflicts affecting major projects should also be considered.

15. The construction period for major projects is relatively short compared with the preparation of the civil engineering scheme. For example motorways are now in preparation for something like 15 years and during this period the design team interfaces with many separate organizations. At present about forty organizations are consulted and wish to share in influencing decisions. Inquiries indeed provide a stage for the conflict to take place formally.

16. What contribution does the Author of Paper 9 feel behavioural sciences could make to reduce the period of planning? Substantial benefits would occur given that the design chosen is the most appropriate.

17. The Authors take different views on systems of management. Paper 9 supports a more direct system whereas Paper 10 advocates a hierarchical organization. What is the evidence or at least sources of evidence for either? Is it not that both apply to different circumstances?

18. There are many areas concerned with major projects open to research of the kind described in Paper 9. For example, we are all aware of the behavioural difficulties which stem from administrative arrangements and perhaps the most obvious example is the way in which contracts are organized between employer, contractor and engineer. This arrangement has evolved over time to deal with conflict and while it has inherited wisdom it has shortcomings of its own which may prove amenable to improvement from a study of the behaviour involved. For example resources on site have escalated merely to deal with measurement. What is the underlying behaviour? Would it be modified by instant arbitration? (Here both the engineer and the contractor put forward their amounts as a fair settlement of a claim and an instant arbiter chooses one of them.)

19. However, the behaviour which has lead to escalation may stem directly from the choice of the lowest tender. It is only possible to speculate what would happen if the tender nearest to the mean less two standard deviations were chosen. Has behavioural science any benefits to offer us in this field of study?

Mr T.D. Kershaw *(Consultant)*

20. Plant and equipment resources, mobilized for a major

overseas project, can at present day prices be valued at several hundred million dollars and equate to more than 15% of the contract value. Consequently, these resources are treated with due care and any person endangering them is soon dealt with. However, the human resources, mobilized for a similar sized job, cannot be valued in money terms, and yet damage to them may only result in a reprimand.

21. When the legacy of an overseas construction project in a developing country can be the skills of indigenous people who have been trained on the project, the need for care and understanding of human resources is even further underlined. Paper 9 gives a general indication of a potential source of help in this matter. However, the unique aspect of major construction contracts needs to be studied more closely, so that this aspect of management can be developed further.

22. The following points are important with regard to human resources.

(a) Senior positions on overseas construction teams should only be filled by individuals with prior overseas experience, albeit at a more junior level.
(b) It is in the best interests of the project that an appreciable proportion of the consultant's field supervisory team, particularly at inspector and junior engineer level, have basic construction experience.
(c) Timely completion of a major project within budget cost is virtually impossible if consultant and contractor teams do not earn each other's mutual respect and work in harmony.

Mr J.G. Broome *(Resident Engineer, A & P Appledore (London) Limited)*

23. The following is a suggested approach to the preparation of industrial relations (IR) on large capital projects in the UK.

24. With reference to union membership on a 'brown field site', it is important to ensure that there is a compatibility between unions which are already represented on site and those which are proposed to be represented during the course of construction. In the early stages the decision should be made as to whether or not

a closed shop arrangement will prevail on site. There are pros and cons in restricting labour in this way but on balance it is probably to the advantage of both management and unions that a closed shop exists on site, being a method of achieving a degree of control over the recruitment of new labour. On the question of union membership, recognition should also be made of the traditional demarcations that exist not only within the union itself but also between tradesmen of different unions. These should be accommodated wherever possible in order to avoid unnecessary conflict.

25. On a multi-disciplined project it is a matter of conjecture whether or not a parity agreement covering both rates of pay and conditions of working is beneficial. On the Cammell Laird Reconstruction Project on Merseyside, parity pay rates existed between some groups of operatives but not others. One of the major factors in determining the suitability of such a parity arrangement seems to be in establishing the degree of acceptance by the parties concerned. For instance, by tradition some working groups have strong branch agreements which would normally tend to override any potential site agreement. However, certain groups of lower paid operatives would have a big incentive to raise their rates of pay to the 'site rate'. One could therefore conclude that parity can only be negotiated to the highest common denominator to the point where it can be argued that a specific group of operatives are exempt by virtue of their own special conditions. Whatever arrangements are made on site it would certainly be disadvantageous to have large variations in rates of pay.

26. These days, most general conditions of working are covered in the UK by legislation, and for different groups of operatives these conditions are largely similar. Special consideration should however be given to arrangements for inclement weather conditions. It is sometimes a source of aggravation on site when certain groups of operatives are exempt from working during bad weather whilst others by virtue of their traditional practices have to soldier on. There are great difficulties in standardizing the working conditions of different groups of operatives which historically have developed their own working patterns. Nevertheless, any standardization which can be achieved will help to minimize inclement weather problems.

27. A method for helping to avoid disputes which appeared to work well on the Cammell Laird Site was to hold regular IR

meetings. These included meetings once a month amongst the operatives themselves to discuss labour and union related matters. There were also monthly management meetings to co-ordinate the IR policies of all contractors on site. Finally, there were integrated monthly meetings between management, union officials and shop stewards to resolve on-going differences. In addition to these regular meetings there were of course many impromptu meetings to resolve particular problems as they arose.

28. From the Employer's point of view a further pre-requisite for setting up the site should be the employment of a contractor experienced in working in the area in question. This is particularly important where IR matters are sensitive because it is the experienced man who has the better chance of achieving success. Not only is it to the benefit of the Employer but also to the Contractor to employ a project manager who is well versed in local working traditions. This also applies to some extent to his staff who, with experience, can recognize the signs of danger and act accordingly. It is the experienced contractor who also has a pool of hourly paid employees who have been with the company over a long period of time. Several benefits accrue from this arrangement which results in increased loyalty to the company who by implication are likely to remain in the area and provide continuity of employment. Encouraging labour in this manner tends to be more efficient (because operatives are familiar with the company's working practices) and less likely to cause industrial disputes (which are often the result of outside influences gaining access to site).

29. Strong official and unofficial contacts between management and unions are essential in avoiding misunderstanding and in promoting rational discussion. They are also useful in providing the necessary feedback on the effect of decisions taken or about to be taken. In short, a good working relationship should be encouraged between all parties concerned in labour related matters.

30. On those projects where IR matters take up a significant proportion of the project manager's time, a site agent or deputy project manager should be employed expressly to execute the Works. This will give the project manager time to consider the implications and possible consequences of alternative moves. It will enable him to consider his strategy and implement his plans in a calmer atmosphere.

Dr Klein

31. Mr Parkes confirmed my hunch that a lot of the research that has been going on in manufacturing industry may not be relevant, that there are motivating aspects in the work already, both in the way gangs work and in the nature of the work itself. He talked about the group dynamics of the people who are working against management. Something is being optimized, and they probably have a very satisfactory system, only it happens to work in a different direction from that of management. I understand his question, but I do not have an answer to it. It is a very big step to understand those dynamics.

32. Coming to Mr Bulman's quotation (paragraph 7), leaving out the 'lazy', 'difficult to control', etc. seems to me the description of a mature, grown-up, well-off human being. I should think most British labour is difficult to control, and probably fussy about living conditions; it is a function of having a rather high standard of living. People who begin to get used to a high standard of living can afford to be fussy and to assert their need to control their own work situation.

33. I do a lot of work in Germany, and on a German railway station I met an English bricklayer who goes to Germany quite often in the summer to get work, and comes back to Manchester in the winter. He complained that one cannot learn German on a German building site because it is full of Italians, Turks, Jugoslavs and so on. In other words, it is not just the British who opt out of such work. I do not know about the 'lazy' — let me avoid that one. However, the 'fussy about living conditions' and demanding not to be totally controlled is a function of Western European society with a high standard of living. The only thing is that other nations have found other ways of dealing with the point. Other nations, too, will get there — which is the point I really wanted to make. Those who are now more docile will not be docile for ever.

34. I thought twice about asking 'Why do you not spend some money on research?' One really does not have instant answers which are better than the experience of people who have been around in a particular field for a long time. I do some work in a bank. That was, for me, the first time in a service industry, as distinct from manufacturing, and when I was asked 'What can you do for us?' I said 'I have no idea. May I have a look round first?'

They were willing to have some research done, after which one did start having ideas. I did not mean to make a plea for research, but if some of these other problems, which are characteristic of the industry, are to be tackled in perhaps a deeper way, research ought to be thought about.

35. The 'hobby horse' which I tentatively put in my paper is this: what about spending a small fraction of the kind of moneys that are allocated to contracts on R & D activities which are about the human and social issues? That could make a dramatic impact.

36. There is one possibility of a suggestion in response to Mr Bridle's question on whether planning time can be shortened. As a citizen, and from reading newspapers, one gets the impression that these planning inquiries happen each time as if there had never been one before, as if it is all new, and a lot of things have to be gone through which might not have to be done every time. Maybe the consultation could be institutionalized in a way which shortens the time taken.

Mr D. Shaw *(General Manager, Construction Industry Training Board)*

37. I do not quite accept that the way to look at problems of the sort mentioned by Mr Lawrenson is to say: 'We must have solutions which are clear-cut and straightforward, like making people answerable by legislation.' Frankly, it does not work. The only way we will get people to work is by winning their co-operation. In the conclusion to Paper 10 it can be seen that in spite of having discussed the various techniques and made various suggestions, Mr Kemp says that these suggestions are simply a framework within which both sides have to operate. The scale of the task is such that management cannot be expected to maintain a constant vigil on every operation. In this situation, the principal objective of management should be to encourage the development of an environment in which the work-force is motivated to control and manage itself. This, I think, also answers Mr Bridle's question (paragraph 17).

11. Running a large contract in the UK

L.C. Allcock, BSc(Hons)*

Introduction

1. When Shell UK Exploration and Production (Shell Expro), who operate in the North Sea on behalf of the Shell and Esso groups of companies, decided to proceed with the development of the South Cormorant offshore oil field in 1973 we were already facing the problem of constructing and installing offshore no fewer than five deep water structures with their associated decks and oil and gas process facilities. The South Cormorant platform was to be the sixth. We had already conducted an evaluation of the few contractors who had the design ability, the onshore construction sites, and the overall management, organization and financial resources to execute such large contracts, both in the UK and on the continent, and we had already made the decision that we would select a concrete structure.

2. At that time the steel fabrication sites in the UK at Nigg Bay, Methil and Graythorpe were already fully occupied and the one partially developed site for concrete structure fabrication at Ardyne Point was already committed to the construction of the Brent 'C' structure (Fig. 1). On the continent also, yards were extensively committed and careful selection of contractor was of critical importance.

3. One of the principal contractors in this field of construction was the Sea Tank Co., a major French engineering design organization associated in joint venture with Sir Robert McAlpine and Sons (McAlpine-Sea Tank) who were the owners of the Ardyne Point site. This venture already had a contract from Shell Expro for the Brent 'C' platform. The contract for Cormorant 'A' (Fig.2)

*Central Engineering Manager, Shell UK Exploration and Production

Design by	Sea Tank Co.
Built by	Sir Robert McAlpine
Water depth	141 m LAT
Maximum No. of wells	40
Storage capacity	550 000 bbls
Design flow capacity	150 000 b/d
Deck area	4000 m^2
Weight of steel in deck	4300 t
Weight of substructure	275 000 t

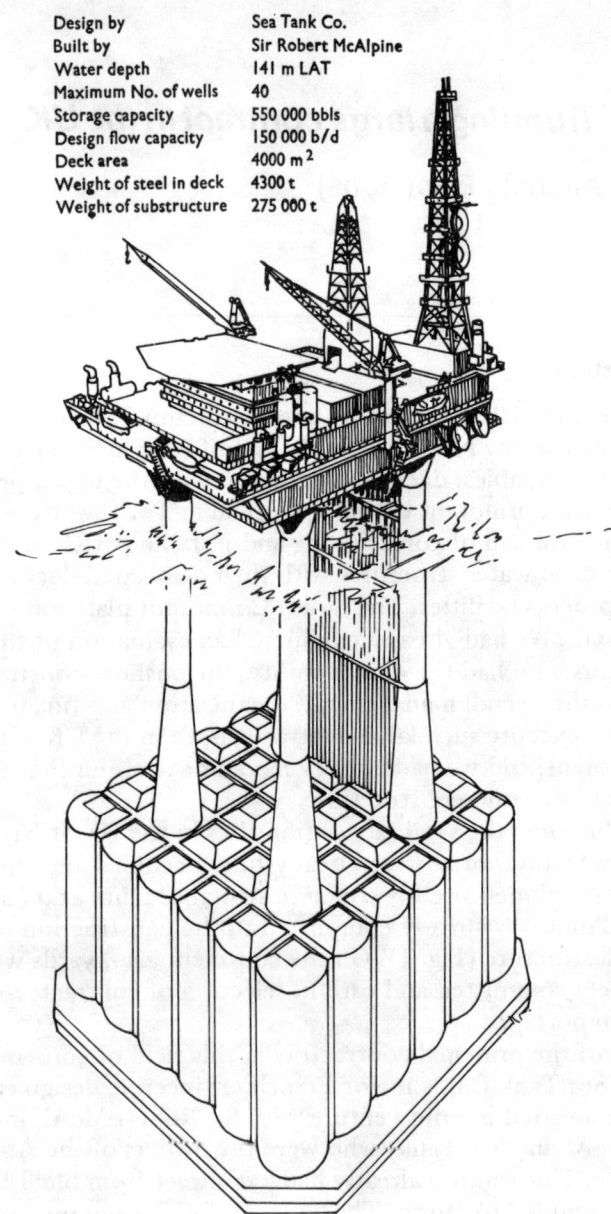

Fig. 1. Brent 'C' platform

differed in detail from the one for Brent 'C' but the major question was the capability of this contractor to handle the two major construction jobs at the same time. It should be noted that the Brent 'C' work had been awarded in December 1973 and Cormorant 'A' was to be awarded in April 1974. The requested delivery date for the second platform was to be one year later than the first, that is 1975 for the former and 1976 for the latter.

4. After lengthy discussion followed by evaluation of the planning put forward by the competing contractors, it was decided to select McAlpine-Sea Tank to construct the Cormorant 'A' structure and deck, and the award was announced in May 1974.

The work

5. To put into context what is called, in the title of this Paper, 'a large contract', it is necessary to spend a little time explaining the magnitude of the work. McAlpine-Sea Tank were contracted to design, supply materials and construct two major structures for installation in the northern North Sea.

6. Brent 'C' was to be placed in 141m of water and, to allow for passage of all expectable waves through the structure, the deck was to be placed an additional 23.5m above sea level: a total height of 164.5m between sea bed and underside of deck. The substructure, without deck, would weigh nearly 300 000t, the caissons at the base would cover an area of nearly 10 500m² and the total superimposed dead load of deck and equipment would weigh over 20 000t. When converted into live load equivalents, with the added weights of an active oil well drilling rig and the weights of produced fluid flowing through the process plants, the figure of superimposed load became of the order of 30 000t. Within the caissons it was also proposed to have storage space for some 550 000 barrels of crude oil.

7. The Cormorant 'A' structure was to be placed in 149m of water with a total height between sea bed and deck of 173m. The substructure, without the deck, would also weigh nearly 300 000t and the sea bed area of the structure would also be some 10 000m². Superimposed loads would be a few thousand tons lower than for Brent 'C' principally because the oil production

173

expected from the Cormorant South field would be of the order of 50% of the Brent 'C' expectation. The storage capacity in the Cormorant 'A' caissons was to be about one million barrels or some $75\,000\text{m}^3$.

8. The cold figures are in themselves quite impressive, but the figures do not represent the totality of the work. Two dry docks were required onshore; after the first stage of construction the bases would be floated in the dry docks, and then towed to an onshore site with prepared moorings capable of withstanding the 100 year return storm condition at Ardyne Point, a site where wind conditions particularly, can be very intense. A bridge had to be constructed and arrangements made to place concrete at the inshore site. Concrete pouring was programmed on a 24 hour basis using a sliding formwork system to construct the four columns. It is of course obvious, although easily overlooked, that the structure would be floating at that time and would have to be maintained in a marine stable condition throughout the construction by a combination of increasing ballast and draught in a controlled manner.

Design

9. Massive concrete structures of this type, for use as permanent offshore islands on which a major industry could carry forward its activities, were relatively unknown. The fact that the industrial activity involved drilling for high pressure oil and gas, and safety handling these hazardous fluids, caused an added area of problems.

10. Primarily, however, the Contractors were called upon to design structures for conditions of very high dynamic loadings in a severe marine environment. The structures were not the first in their field, nevertheless no designer in the world in 1973 had climbed far up the learning curve. The UK Government had, over the same period, introduced legislation which required the oil industry to have designs of offshore installations checked by one of five outside agencies, on whose recommendation the Government, through the Department of Energy, would issue to the specific oil company a Certificate of Fitness without which the installation could not be used.

11. Marine classification agencies were appointed whose experience in the field of concrete and fixed offshore installations was very

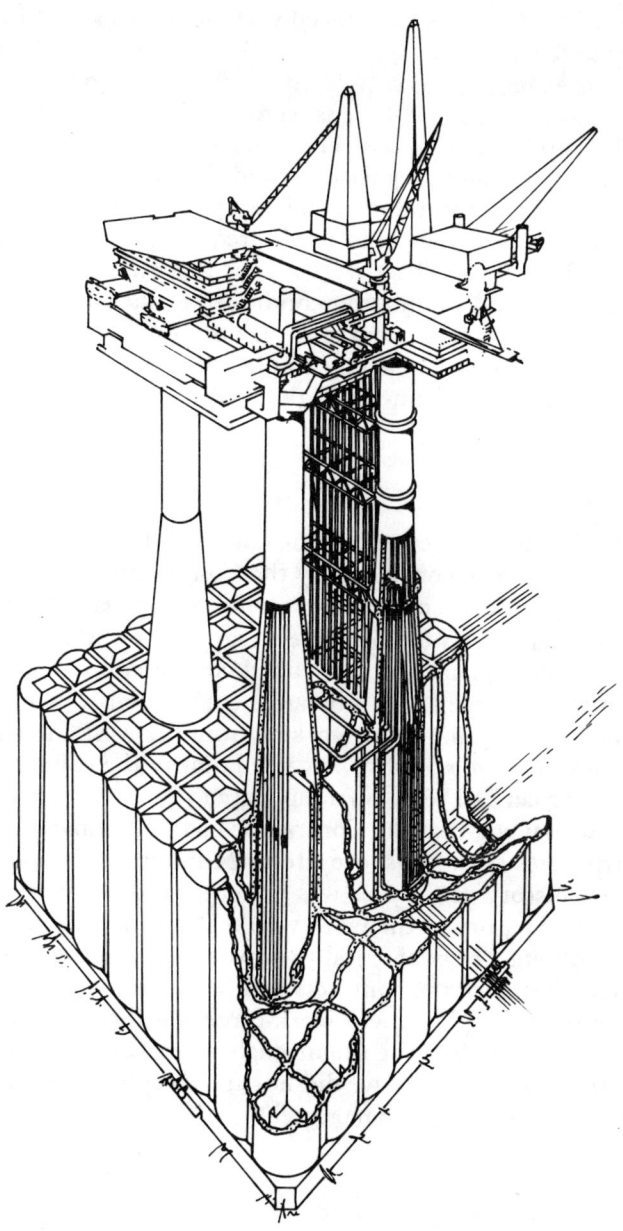

Fig. 2. Cormorant 'A'

limited. The designers were faced with designing structures for poorly understood but severe conditions of service. The design wave, for example, was stipulated as 30.5m for Brent 'C' and 29.8m for Cormorant 'A'. The designs were required to conform with slightly conflicting Codes of Practice and guidelines newly established by government regulations, and therefore unsubstantiated by precedent interpretations, and also to be flexible enough to accommodate for changes in service requirements which the Client was still firming up. There were also at that stage rather less than definitive soil and foundation conditions at the sites. As a starting point it was not very encouraging, if only for the fact that although preliminary designs had been completed, no one could foresee what problem areas would emerge as detailed design progressed.

Work-force

12. Within the framework of the contract the Client was also interested in being convinced by the Contractor that a sufficient and skilful work-force could be mobilized to execute the work at Ardyne Point. It would be naive to say that we were fully convinced, but the plans which included construction of a large construction village at the site and a comprehensive scheme to commute large numbers of workers from the west Glasgow area and from the towns west of Ardyne Point in the county of Argyll by road and ferry boat, appeared sound. The main source of our worry was that the Contractor had undertaken to deliver the structures at the offshore site in 24 months. The industrial relations history of the area was not good and we believed that the unusual nature of the work would be fertile ground for work-force problems. Fundamentally, however, we as clients adopted the view that once the contracts had been awarded the relationship between Employer and work-force was the Contractor's business and much as the Client might worry about strikes and work-force productivity we did not feel deeply involved in that aspect of the contract management.

Form and value of the contract

13. We had issued tender documents to the contract industry on

the premise that the design was sufficiently advanced and the work sufficiently defined to permit a lump sum form of contract. Allowance was made in the price structure for escalation by tying in with the General Conditions of Contract, fourth edition published by the Institution of Civil Engineers. Various other more specific indices were to be applied to particular aspects of the work. We had also allowed for and agreed that after completion of the works at Ardyne Point, the towing of the structures to a deep water site, placement of the decks, the second stage of towing to the offshore site and the final grouting and outfitting would be executed on a reimbursable cost basis.

14. The contract industry were prepared to tender competitively for the job and the tenders were evaluated on the basis outlined. The original contracts, determined in 1973 and early 1974, were valued respectively at £24 million for Brent 'C' and £33 million for Cormorant 'A'. A significant element of these estimated values was to be paid in french francs and another smaller element in dutch guilders. What we did not know was the impact on these estimated values of future design changes, the delays to be occasioned by labour unrest, strikes and poor productivity, the severity of escalation which was to bite so deeply after the OPEC increase in crude oil prices was imposed during 1973, and the sliding devaluation of the pound sterling against the continental currencies.

15. Perhaps the worst feature, which was an effect rather than a cause, was that delivery from Ardyne Point did not take place until mid 1977. This was late completion by some 18 months in the one case and 15 months in the other for the respective structures. This meant that escalation took place over a greatly extended period. Over the course of the contract the effect of escalation as measured by the relevant indices, added to the effect of sterling devaluation, increased costs by some 125%.

16. The final costs to the Client, excluding any financial effects of delayed return on investment, which were agreed after prolonged investigation and detailed substantiation of actual costs and a great deal of compromise, amounted to some £75 million for Brent ·'C' and nearly £115 million for Cormorant 'A'. Against original estimates of £57 million for the contracts up to the point of delivery of the structures from the Ardyne Point moorings, we paid the Contractor nearly £190 million. This was slightly more

than triple the cost. If one theoretically allows for 125% escalation, the £190 million would reduce to £84 million (1973 costs) and the increases would then appear far less appalling.

17. The Contracts also specified the terms and conditions of payment to the Contractor. In contractual terms the initial phases were not normal in the sense that although the Clients were to be the future owners of the structure, we were not the owners of the construction site. The Contract called for the Client to pay the Contractor certain sums as preliminary payments for conversion of a green field site into a construction yard and thereafter to make monthly payments against progress. The progress was to be measured in terms of work completed and measured against a previously agreed schedule. I should say that the agreement to pay preliminaries for a contract at the contractor's own site, before any tangible asset is created does not appeal to us and occurred in this case largely because of the high cost of the major site preparation work which the Contractor had to carry out at Ardyne Point before major construction work could be started.

Geography of the contract

18. Part of the complexity of this particular major contract was occasioned by the rather unusual geography associated with it. The concrete structures were to be built at Ardyne Point on the Western side of the mouth of the River Clyde in Scotland. The deck for one structure was partially prefabricated at Methil on the Firth of Forth and fabricated in the Netherlands. The deck for the other structure was fabricated near Marseilles in the south of France. The packages of prefabricated pipe work to be installed in one of the legs of Cormorant 'A' were assembled on Clydeside whereas for Brent 'C' the tubulars in the so-called utility leg were fabricated and installed at Ardyne Point itself. The designs for the structures were made by Sea Tank Co. in Paris and the designs of the site works were made in London by McAlpine.

19. The Client's main office administering the whole of the Ardyne Point activity and associated works was in London with site project management teams at the various sites. The site management team at Ardyne was mobilized by Shaw and Hatton Consultants Limited. Under the regulations of the Govern-

ment a classification society had to be appointed and the Client chose Det Norske Veritas (DNV) to carry out this work. We made this appointment primarily because DNV had already built up a history of experience with other major concrete structures which were under construction in Norway.

20. We as the Client were not well staffed with respect to experienced concrete design engineers and we felt particularly concerned at the quality of engineering expertise in soils and foundation engineering available to appraise the designs of the foundations. To assist us in this matter of design audit we engaged the Instituut TNO voor Bouwmaterialen en Bouwconstructies, a highly technical organization who had a long experience in the design of the Delta works in Holland and whose expertise in concrete design and the evaluation of the foundation stability of large concrete structures enjoyed an international reputation. TNO were established near The Hague

21. The management of the contract was made much more complex by this wide spread of responsibility. Designs were turned into drawings and technical explanation by French designers in Paris. This data had to be checked generally by our own engineers in London and thereafter in detail by DNV in Oslo and TNO in The Hague. Whenever a query arose the back checking and resolution of the difficulty was time consuming and not easy to expedite. There were occasions when design was barely abreast of the site construction and at times emotion was close to the surface.

Management of the contracts

22. It has seemed necessary to explain at some length the complex background of these major contracts in order to explain the way in which we endeavoured to manage them. We were faced with a shortage of experienced staff, and in London we were organized to gain whatever advantage we could from the experience we had available. The commitment was to construct six major platforms, five of which were concrete gravity structures. We therefore concentrated our structural design experience into one department within our engineering division. In this way we developed a better understanding of the design problems of the concrete structure and were able to build something of a person to person relation-

ship with DNV and TNO. This, we were confident, was a better approach than setting up separate project teams for each structure, each with their own design expertise.

23. It may well be asked why we felt obliged to expend so much effort on design validification but the answer is relatively simple. These designs were for relatively prototype structures and we would have felt insecure if we had not become directly involved. Shell Exploration and Production is a company operating in the North Sea on behalf of the Shell and Esso groups. Our parent companies are very strong in engineering expertise and we knew that we would be able to call on involvement of a range of experienced specialists should specific problems occur. In the event, this strong involvement in the design of the two structures was a major contributory factor to the successful outcome of the project. Many well known specialists from other organizations participated in resolving the problems of the foundation and structural design and we were encouraged to believe that but for this involvement the many problems would have taken much longer to resolve with consequent further delay of the work.

24. On the construction side of the contracts, we built up a team of project engineers who were delegated the responsibility of ensuring that the construction work progressed on time and within the budget. As already explained and largely for reasons outside their control the work was significantly delayed and costs increased very considerably, but nevertheless the responsibility of the project teams was there.

25. In London we appointed a Head of Project who was specifically in charge of the Ardyne Point contract and for construction of the off site works associated with Brent 'C' and Cormorant 'A'. Under his control, also in London, we appointed two small teams of engineers with planning engineers, an accountant and a materials supervisor, each team with a responsibility for a single platform. In Ardyne Point we also set up a small site team. The site project management was contracted out to Shaw and Hatton Consultants Limited and as general superintendents of our interest we appointed a Shell Expro resident engineer and a deputy.

26. At Ardyne Point, in London and in Paris, regular meetings took place between Contractor and Client to discuss progress and difficulties and, of course, even though the Shell Expro design specialists were not directly responsible to the Head of Project,

they were deeply integrated functionally into the project management teams. Responsibility for payment of invoices was also
within the scope of the project management team who were
required to validate all requests for payment. The payments themselves and the creation of the finance structure to make payments
was not under the direct control of the project team. Shell Expro
has a large finance and accounts division who work in close relationship with the financial divisions of the affiliate companies. It was
their responsibility to ensure that the money was transferred to
the contractor's accounts once the invoices had been authenti-
cated by the engineers. I should also mention that teams of
engineers similar to the one at Ardyne Point were also appointed
to the deck construction sites.

27. Part of the project management responsibility was the
measurement of progress on the site. This was required under the
contract so that monthly payments against measured progress
could be made. This monthly measurement and value calculation
was carried out by quantity surveyors at the site and the measure-
ments were agreed in discussion with the contractor's own quantity
surveying department.

28. This was the basic organization for management of the contract
but, as seems inevitable in the UK at present, and in the light of
the design changes that had to be made and the relatively poor
productivity at the site, particularly in the first year or so of the
contract term, large claims began to emerge which became of
such great importance to the progress of the work that special
additional staff had to be mobilized to investigate and validate
these claims. The lump sum nature of the contract with allowance
for escalation was unfortunately not the best basis for the negotiation and substantiation of contractors' claims. The onus of proof
lay with the Contractor but substantiation was not easy. Productivity was running lower than expected and there was a shortfall between progress payments and the contractor's expenditures
and to continue with the work meant that the Client had to
recognize this problem.

29. Design changes were another source of difficulty. It was
relatively simple to agree that a design change represented an
additional direct cost to the Contractor but it was not simple to
concur with him on the added cost of disruption to the active
contracted work that he had planned and costed. To understand

and agree the value of the contractor's claims required the Client to carry out a major special investigation into the Contractor's project expenditures. The team appointed to do this was made up principally of accountants and engineers who had specialized in cost engineering. The Clients, I might say, were greatly troubled by the need for this and the time required to carry out the work involved. We were, however, able to respond to the added load and to maintain the progress of the contract.

30. The poor progress and the many problems of the first two years of these major works were a phase that passed. The productivity and progress achieved in the last eighteen months of the work programme at Ardyne were most impressive and although it was obviously impossible to recover the time lost in the early stages the Client was most gratified by the final progress achieved.

Conclusion

31. If we were to repeat the experience of these major contracts we would probably not proceed in quite the same way. Far more complete designs would probably be insisted on before contracts were awarded. We would also require a plan and schedule of the construction in much greater detail and involving the use of the best planning and costing techniques much more extensively. We would also do our utmost to prevent industrial unrest and would become much more involved at an earlier stage. All this is hindsight, but on the whole the contract work at Ardyne was a success, only troubled by factors beyond control of both Client and Contractor.

12. Managing an overseas contract in a developed country – Bosporus Bridge

D.C.C. Dixon, BA, FICE, FInstW *

Introduction

1. Modern Turkey is a country physically divided by the Bosporus. This narrow stretch of water, some 22 km long and just under 1 km wide at its narrowest point, links the Black Sea in the North with the sea of Marmora in the South. About 2500 years ago the Persian armies of Darius and Xerxes crossed the Bosporus on a bridge of boats. It was not, however, until the latter half of the nineteenth century that the natural desire for a permanent crossing was matched by engineering techniques for even tentative engineering solutions. In the early 1950s the development of inter-city road communication led the Government to give serious consideration to a permanent crossing. A traffic study by De Leuw Cather in 1956 concluded that such a crossing was both feasible and economically viable.

2. Political instability in Turkey in 1960 caused delays and it was not until 1967 that, persuaded by the increasing bottlenecks on the ferries, the Government included the bridge[1,2] in its five year programme. Engineering studies were invited from consulting engineers in both the United States and Europe; Freeman Fox and Partners presented their preliminary proposals in October 1967 and a formal agreement was made with them by the Turkish Government in January 1968. The intention was to invite tenders early in 1969 but, due to the difficulty in arranging international financing, it was not possible until June of that year.

3. The train of events leading up to the construction of a major bridge followed the usual pattern — years pass by as the need for a major crossing becomes more evident and the decision and the

*Marketing Director, Cementation Civil, Structural and Offshore Engineering
    Holdings Ltd, and Deputy Chairman, Cleveland Bridge and Engineering Co. Ltd

means to achieve it are found; the time for construction becomes even shorter, the burden of achieving speed by the builder the greater.

4. The design was a suspension bridge with six lanes of traffic having a main span of 1074 m.

Relations between Hochtief and Cleveland

5. Hochtief A.G. and the Cleveland Bridge and Engineering Co. Ltd, agreed early in 1968 to form an association for the construction of the bridge. They intended that other companies should join them but detailed negotiations with several companies as to the form of association led to nothing and, with three weeks to the time of tender, the two companies found themselves alone. Nevertheless they went ahead which shows the extent of mutual trust and respect and of a clear understanding of the duties and responsibilities of each. The agreement was for a consortium; that is to say each company was totally responsible for its own portion of the works — Hochtief for the foundations and the sidespan concrete roadway decks, Cleveland for the steel superstructure — and each for his own profit or loss on these portions and for any liability which might arise. The agreement included a provision that one partner would not claim on the other for consequential costs (of delay by the other for example) and that any liquidated damages would be met in equal shares regardless of cause. Such agreement engendered excellent co-operation and relations throughout the contract.

Finance

6. Loans were available as follows:

	US $ equivalent
Germany	7 500 000
France	3 750 000
Italy	1 875 000
Japan	22 500 000
UK	6 300 000
European Investment Bank	11 250 000

Each currency was converted to US $ at a fixed, national exchange rate and contractors were required to state in their bids the amount of each currency they required. Thus the financial arrangements, with the possible exception of Japan, meant that the contracts for steel fabrication had to be placed in several countries. It also meant that companies tendering had little freedom of action, after the contract was awarded, in where they placed their orders. This was a matter of much concern because there was no escalation clause where suppliers from outside Turkey were concerned and there could be no binding obligation on the supplier due to the impossibility of establishing the equivalence of national standards, qualities of steel etc., before the main contract was placed. Cleveland were further restricted by the fact that their German partner not unnaturally required a considerable portion of the DM available. A late difficulty arose in that no aggregate suitable for the mastic asphalt on the deck could be found in Turkey and all the ingredients had to be shipped from the UK.

7. In the event all the pre-tender offers were converted into firm, fixed price contracts without any significant variation from the tender allowances. Nevertheless a considerable risk had been taken and surmounted; the energy and urgency shown in obtaining these contracts was well rewarded when inflation reached serious levels in 1970 and 1971.

8. Such arrangements also created complex problems for the Contractor who was paying and being paid in five different currencies with differing exchange rates, and where the rates of borrowing or lending in each of the countries concerned were constantly changing. The terms and conditions of the Contract necessitated borrowings and deposits in these countries. Special computer programmes were devised to indicate the ever changing position. The clarity and to some extent the validity of the analysis of the situation at any one moment was doubtful.

Tenders and negotiations

9. Tenders received, adjusted for alternative methods and materials as permitted by the contract documents, were

	$
Krupp/MAN/Redpath Dorman Long	36 665 000
IHI/Julius Berger	36 031 000

Beton and Monierbau/Sir William Arrol   35 427 000
Hochtief/Cleveland Bridge and Engineering Co.   33 721 000

A significant saving offered was by the Hochtief/Cleveland consortium who proposed to use the conventional in situ aerial cable wire 'spinning' procedure instead of preforming the wires in strands. In any case they had been unable to convince themselves that a system previously only used on the Narrangasset Bridge of 488 m span could be extrapolated to a bridge of 1074 m without, at least, extensive teething troubles.

10. All tenders submitted had reservations; the covering letter of Hochtief/Cleveland contained 27. There were five months of negotiations, mostly in Ankara but partly with the European Investment Bank in Luxembourg, before a contract could be placed. The principle issue was the amount of working capital required to finance the construction. On the worst interpretation of the documents Cleveland would have had to find at one particular point in time nearly 45% of the value of its bill items. Their negotiator was permitted a limit of 10%. Advance payments were provided for but their use was at the discretion of the Turkish Government; the requirement for the payment of customs dues was not clear. The outcome of these protracted negotiations was a clear Contract and a cash flow acceptable to the Contractors. The documents required the provision of a bid bond payable on demand; technically this was at risk throughout the five months.

Programme

11. The contract period was 1020 days; less than three years compared with periods of three and a half to six years on comparable bridges in the USA or the UK which were not in a country overseas from the Contractor. Some of this saving was due to the relatively easier pier and anchorage construction but this meant that the early fabrication and the very large amount of engineering, design and manufacture had to be done in about one year less than had usually been available — and needed. A suspension bridge is extremely demanding in the number of separate special pieces of permanent and temporary material required and in the very precise order and date at which each is required. Add to this the fact that they had to be

purchased and shipped, in the event from 16 different ports in
several countries, and that little space was available at Site for
stockpiling, and it can be readily seen what a major problem of
logistics existed.

12. A special organization was set up at Darlington, controlled by
engineers who alone, from their knowledge of the design, specifications and the construction programme, could determine what was
wanted and what alternatives were permissible. A computer program was initiated at Hochtief as required by the contract but the
effort of adapting the program to the very specialist circumstances
and the task of comprehending and making use of the results, to
some extent invalidated the benefits. Hard and systematic work to
control procurement, shipping and delivery was the main tool. During
the early stages of the contract Cementation, Cleveland's parent
company, was acquired by Trafalgar House. The specialist services
of such a major group which could be called upon at any time
proved of great value.

Execution

13. The construction of the superstructure of a suspension bridge
requires an exceptional amount of off-site engineering and this was
carried out at Darlington. At Site the labour employed on site
construction was entirely Turkish. The British staff on site numbered
20 including administrative staff, and the Turkish numbered at
some times over 400 since there were no major problems in finding
or training the necessary skills. In particular the output and quality
of welders was extremely satisfactory and the efficiencies achieved
in the complicated process of cable spinning were at least equal to
previous standards in other countries.

14. As is standard practice in Turkey a 'bargaining contract' was
agreed before work started with one of the major construction
unions and all labour of whatever craft belonged to this one union.
The terms of the Contract were fully observed and throughout
the superstructure construction there were no stoppages of work
caused by disputes. Legislation concerning employment and its
termination is strict and the necessity of thoroughness in investigation of it during the tender period must be emphasized. For
example Tax and Social Security Clearance had to be obtained

not only for men employed by us but also for all those who had been employed by sub-contractors before release of the performance bond could be obtained.

15. Regulations for the issue of import licences and for the handling of supplies, many of great weight, were complicated. A problem existed in pursuading engineers, busy with construction, to direct their time to motivating officialdom to expedite clearance of these goods. A local condition which had to be constantly borne in mind whilst planning was the lack of plant hire facilities. In advanced countries it is normal practice to rely on hired heavy mobile cranes particularly in the initial erection of equipment; in a case such as this every conceivable lift has to be planned. All plant and temporary materials entered the country on Temporary Import Licences and as such had to be re-exported. There was considerable difficulty and cost in 'accounting' for all the myriad of items.

16. Hazards abound in the execution of such a contract. Not least but most spectacular was a storm on 26 February, 1973, in the Marmora Sea which sent unprecedented waves up the Bosporus, and boxes each weighing up to 40 t were swept off the quay into 10 m of water!

Expatriate staff

17. The writing of contracts for staff on overseas contracts which provide reasonable and equitable remuneration and conditions is always a problem, and Bosporus proved no exception. The basis of assessing remuneration was that, for men (with or without their families) living to a reasonable standard in Istanbul, each should be able to remit some appropriate sum as savings to the UK. However, there were many problems. If inflation at varying rates occurred in both the UK and Turkey, what adjustments should be made? (Employees in the UK were limited as to their increases.) Educational facilities were available in Istanbul only up to the age of 11, so should different allowances be made to those with families of different ages and sizes? (Those qualified to do the engineering were very few.) Bachelors could well feel that it was not equitable that their remuneration should be unfavourable compared with those who were married and had families. There is of course no

perfect answer to such problems, the best degree of equity can be obtained by proper consideration of the problems at the inception of the contract.

Standards of materials and workmanship

18. The bridge and its component parts were designed to British Standards with special overriding requirements where necessary. The supply contracts were let all over Europe. Thus the equivalence of standards, the availability of certain steel sections and the acceptability of materials and methods, all to be approved by the Engineer, were major preoccupations in the early days of the contract. Were it not possible to place an order in a particular country as planned then the consequences could be major and complex in that the limitations of currency might make it necessary to change the source of other items of supply. In this case inevitably there were considerable problems of inspection required by both the Engineer and the Contractor.

19. An example of the flexibility required lay in the fabrication of the two towers, made at two separate Italian fabricators at separate works. They were made by different methods but each had to achieve the same exacting results in dimensional accuracy; this accuracy produced a major benefit when one panel was wrecked by careless handling in the port of Istanbul. A replacement panel was made working from theoretical dimensions with holes drilled full size. It fitted perfectly and a considerable delay was avoided.

Costing

20. Some 70% of the value of the work carried out by Cleveland consisted of supply items for which firm bids had been obtained. In these cases it was therefore variations and claims only which required monitoring.

21. Standard methods were used in the control of costs of the site operations. The main difficulties occurred in shipping where global allowances had been used to cover many items, the general overhead items which had been similarly dealt with, and in assessment of the situation. The position was continually changing

Fig. 1. Bosporus Bridge

due to changes in exchange rates etc., and therefore the cost of financing the operation was affected.

22. In practice separate accounts for all currencies other than Turkish Lira were kept in the UK and those for Turkish Lira in Turkey. Consolidation was achieved by using set rates of exchange between each currency, and profits and losses due to transfers from one currency into another at the ruling rate of exchange were taken into the profit and loss account as and when they were made. By the time completion was nearly achieved, provisions had been made to ensure that there was no apparent sudden change in the final position.

Conclusion

23. This was a major contract of considerable technical difficulty, completed successfully in 1180 days (see Fig. 1), the delay of 160 days being largely due to differences in ground conditions in the main pier foundations. In the Author's opinion the most important factor in this success lay in the unified control. There were only two partners: one responsible for all the civil works, the other for all the steelwork. Their joint meetings concerned mutual interests with the Client, the Consulting Engineer, their programmes, etc., but the actual direction of the work lay directly with each of them. Had there been further partners rather than sub-contractors then it would have been necessary to have organization to agree all details of the work and this would have blunted the decisiveness that comes from single control, and which was so necessary on this contract.

24. If there is a lesson to be re-learned, it is that the reconnaissance of the site, and the study of the legislation, conditions and practices of an overseas country during the pre-tender period cannot be too thorough.

References

1 BROWN W.C. and PARSONS M.F. Bosporus Bridge Part 1: History of design. *Proc. Instn Civ. Engrs*, Part 1, 1975, Vol. 58, Nov., 505–532
2 KNOX H.S.G. Bosporus Bridge Part 2: Construction of superstructure. *Proc. Instn Civ. Engrs*, Part 1, 1975, Vol. 58, Nov., 533–567

13.   Managing a contract in a developing country

J.M. Thomas, ACGI, FICE*

Introduction

1. In presenting this Paper it is necessary to define the term 'developing country'. All political entities can be said to be developing, but at different rates. The Paper restricts itself to countries in which the infrastructure and basic economy are still being developed and because of historical circumstances is not relevant to much of Europe and North America. This takes us therefore to parts of the world in which the number and experience in the carrying out of large capital projects is limited, and as a result the British manager is faced with a series of circumstances which are radically different to those in the UK for instance. These differences fall into two main categories: the physical environment of remote areas in which developments are being carried out, and the human, political and commercial factors associated with working in a country with a labour force and an Employer having a different background of experience, and often different priorities.

Factors relating to the physical environment

2. With regard to remoteness of sites and extreme climatic conditions the following important points have to be well taken care of.

Site investigation

3. A thorough review of the available physical data is essential.

---

* Joint Deputy Managing Director, Taylor Woodrow International Ltd

In the UK there is a body of experience of working in most types of materials, and solutions have been developed to overcome difficulties. In developing countries this background will probably be absent and the behaviour of for instance tropical decomposed rocks and their derived soils will require consideration. In the same way data on meteorology, river flows and run-offs may well be sparse and unreliable, and the design of temporary works must provide the right answer on sound engineering judgement. Maps may well be unavailable and the decision on the required standard of accuracy for surveys may have to be decided.

Materials

4. The constructor must find building materials of the required quality which will involve the investigation of quarries and the obtaining of fine aggregates from a variety of possible sources. Testing will have to be carried out to assure strength properties and the absence of deleterious substances. In some areas the lack of fresh water may require the mixing of concrete with sea water and certainly its general use for curing and cleaning. This introduces a range of problems.

Contract planning

5. In works carried out in a developing country there is very little opportunity to change construction methods, once chosen quickly, by for instance drawing on a pool of experienced sub-contractors, as would be the case in the UK. The method of construction decided on therefore has to be right, and for this reason well proven methods suitable for the standard of labour available have to be chosen meticulously. The ordering and delivering of materials must be clearly scheduled to allow contingency time for shipping delays. Arrangements have to be made for the good storage of materials bearing in mind that an omission or shortage may require expensive air freighting of the missing elements.

6. The construction equipment must be chosen not only for the principal construction demands, but also if possible to have sufficient flexibility to carry out ancillary tasks such as plant erection, and to provide back-up to other equipment. To reduce

the stocks of spare parts and to give inter-changeability the equipment should be standardized to the greatest extent. If possible the same power units should be used for compressors and other small plant, which will ease both the spare parts situation and also maintenance and training. A fully equipped engineering workshop will be necessary and the rebuilding of spare parts will be a feature of maintenance practice.

Contract organization

7. In order to carry out a large project in a short construction period it is obviously necessary to have a strong management organization and the right balance between the decisions taken at Head Office and those delegated to site level. The view on this varies among major contractors, and it is also affected by the facility of communication between the two offices. On very large contracts sophisticated data links can be justified, but on all contracts telephone and telex services, as a supplement to airmail and air courier, will be normal.

8. Visits from directors and senior responsible managers are necessary on a regular basis because however competent the site management may be, the broader view which can distinguish the wood from the trees is essential. Management must be careful not to allow overseas visits to become only trouble-shooting expeditions tied to current urgent problems. A visit to Site at an earlier stage can possibly forestall the build-up of a problem which might commit large amounts of senior management time if left. An essential feature of these visits is the morale building and maintenance of the site staff by becoming familiar with their day-to-day conditions and difficulties. A further commitment of senior management time will be to maintain good relations with the Employers. In developing countries this is frequently on a personal basis, and visits by more junior staff could be looked upon as an affront.

9. Contract procedures must be worked out in complete detail. The paperwork and the timing of an application for payment, with possibly complicated financial arrangements requiring authorized signatures can be a management achievement; the construction is relatively simple in comparison. An experienced company will have developed a range of standard procedures

195

adjusted for a particular contract, and these will form a Contract Manual which, together with the support of strong experienced service departments for buying, shipping and personnel, can relieve line management of much detailed work and allow them to concentrate on steering the ship.

Engineering design

10. In many developing countries it is increasingly becoming the practice for the Contractor to be responsible for the production of the design. This will require the employment of experienced conceptual designers with their support teams to produce work to the highest standard in a very short time, as the design can only be started after the award of the contract. This produces pressures which are very different to those for a contract in which a tender is invited on the basis of full working drawings, which may have been prepared over a period of years. It will require a good knowledge of acceptable design codes and a familiarity with earthquake loadings and other similar local phenomena.

11. It is essential that the engineering not only provides the most proven economic solution, but that working drawings are produced to a programme which will give sufficient lead-in time for the detailing to include all built-in items and their procurement on to Site. The construction procedures embodied in the designs should be as simple as possible bearing in mind that they may well be constructed by semi-skilled labour under accelerated construction times. There are advantages therefore in using more material if this simplifies the structure and reduces complicated.shuttering and concreting sequences.

12. It is part of the British civil engineering tradition to produce high quality designs, but this tradition is based on craftsmanship which is becoming more and more difficult to obtain in the UK, let alone in most developing countries. The design should embody the construction practices most suited to the deployment of large capacity construction equipment. It is by this means that fast construction programmes can be met. Therefore if the management has control of the design process it should ensure the use of value engineering judgements, and in this respect the standardization of detailing developed in the USA, which is further removed from the craftsman era, can be of use.

Site facilities and accommodation

13. It is frequently necessary to provide complete accommodation for expatriate staff and labour, which requires the building of a small township with full services including a school and hospital. The Project Manager in the developing country will quickly find himself involved in family welfare and health care extending from the cradle to the grave.

14. Off-loading facilities for ships may well be required, and the laying of mooring buoys and the provision of jetties, lighters and tugs. Haulage roads may have to be built or improved, and the assessment of the economics of the quality road against maintenance cost and life of haulage units may be a major financial decision.

Commercial and political environment

15. In many ways a contract for which the manager supplies everything from offshore would appear to be the easiest to administer. However, requirements of economy and the aspirations of the host country in the construction business prevents this, except for a very small number of special projects. The requirement therefore is generally for a degree of association with a local organization which if properly handled, produces extremely beneficial results in local politics and knowledge, language, and administrative and local legal requirements.

Labour

16. On most projects it will either be necessary to recruit and train the local labour or, as is happening in parts of the Middle East, import labour from outside the country. There is considerable discussion on the economics of different kinds of labour. In the past there was a belief that labour costs were self-adjusting and that the higher paid labour produced higher outputs and vice versa. But intervention of modern training techniques has unbalanced this equation and much labour, which in the recent past was considered only suited for unskilled work, is being trained up for plant operating where their stamina and endurance lead to high outputs. Communication with labour is important and British managers

have an advantage in that english is so widely known in the main labour recruitment areas.

Staff

17. Selection of staff is probably the most vital element in the successful completion of contracts and again British managers have an advantage. As a nation we are still relatively outward looking and work overseas is part of our civil engineering tradition.

18. The adjustment of inexperienced staff to primitive conditions and labour can be traumatic and has been referred to by Dr George Foster, Professor of Anthropology at the University of California, as 'culture shock'. He describes a neurosis which affects many engineers placed in an unfamilar overseas environment. It is the dazed condition that occurs when they first see their elaborate specifications used for rolling cigarettes, and realizing that they are not communicating at all. The condition deepens as they become more and more obsessed with doing the job in the 'right way'. Symptoms of culture shock are such things as hand washing obsession and the continued use of the phrase 'these people', i.e. 'These people should appreciate'; 'These people should understand'; 'These people cannot plan'.

19. This problem is most quickly overcome by the leadership from the top, the generation of team spirit and the seasoning of the staff with tried and proven veterens used to such conditions. It is also helped by the encouragement of staff to learn the local language and take an interest in national customs.

20. Changes have occurred in the staff recruitment situation as a result of the lower age of marriage, the influence of the welfare state, and in particular the benefits of what is described as the social wage. In the recent past managers have found that the stability of married staff and the contribution which the ladies have made to social life in isolated projects has offset the higher cost of housing and air fares. Nowadays many of these young ladies have qualifications and careers of their own, and the lack of opportunity for them to practise these in many developing countries may dissuade the couple from working overseas.

21. Developing countries are tending to increase the opportunities for their own construction cadres by restricting entry permits to expatriates with either senior responsibilities or high professional

qualifications. Every contractor knows that the success of his operations in difficult conditions is dependent on his experienced trades foremen, and it is for this category that it is difficult to obtain visas.

The Consulting Engineer

22. The consulting engineer if one is retained by the Employer is critical to the success of the contract. British consultants have a unique position in carrying out major projects in developing countries resulting from their high standards. However, it is with some foreboding that a major British contractor begins a contract which has been awarded on most severe competitive terms; not only will he be expected to provide an adequate job, but indeed to produce a monument to the British civil engineering industry! The higher his reputation and the financial background of his firm, the more money he will be expected to spend for this purpose, although other contractors who customarily work to lower but adequate quality standards are allowed to continue as they are.

Sub-contractors

23. There is often a very natural inclination to keep strict control by employing labour directly. However in some developing countries, particularly in South America, the labour laws are so stringent that sub-contracting is a necessity. Practices to circumvent these laws are normal for indigenous contractors, but would be taken as a disrespect if engaged in by expatriate contractors. Great care is needed when dealing with sub-contractors as they may not fully understand those concepts of work content which British managers are brought up with.

The Employer

24. Employers in developing countries are different in many ways, but generally have in common the expectation that everything should have been done yesterday and for a fixed price. The traditional British practices of variations of price and re-measurement have little appeal, while other nations with different home industries, stronger currencies, lower inflation rates and possibly

tax advantages will agree to the conditions of fixed price and penal bonds. Whether these conditions will change is difficult to say but the problem affects the more substantial British owned firms who do not have the option to go into liquidation in the event of a disastrous combination of unforeseen circumstances; for these companies large areas of the world are becoming increasingly difficult.

25. The Employer may not understand clearly the role of his Consulting Engineer. As he pays the fees he may well feel entitled to the complete support of his Consultant on all matters and have little understanding of the tradition of impartial judgement on which the profession is founded. The settlement of disputes may well be in local courts with local legal codes and these again can produce uncertainties for management.

Conclusion

26. The sole business of the Contractor is to build and to be cheerful in adversity. The Employer in the developing country often does not fully understand nor wish to entertain the detailed and often tenuous contractual claims which are so much a part of the British construction scene. Fifty years ago we could have been able to demonstrate conclusively to developing countries that our practices and methods were best, and at that time they undoubtedly were. We now take to the world market place a domestic situation in which the pressures on management of labour problems and claims, together with a major decline in the kind of projects in which high production is essential, is producing an attitude which requires a great deal of transformation before it can be utilized successfully in a developing country.

27. This would appear a gloomy conclusion if it was not that major British firms do operate successfully, in the conditions described in this Paper, as the result of great experience. Those who are most successful over a period in carrying out a large amount of projects in difficult overseas areas collect procedures and experiences which can stand them in good stead in an increasingly competitive business.

14. Commissioning and handing over of plant in a developing country

A.M. Brown, BAI(TCD), FIMechE*

Introduction

1. It is widespread practice in industrially developed countries for the staff of the owner of plant to become deeply and closely involved in plant operation from the pre-commissioning stage onwards. Since the owner has the task of long term operation and maintenance, this is a prudent course of action from his point of view, enabling his staff to become acquainted with the plant in a simple and direct way. It also allows close contact between the owner's operating staff and the contractor's commissioning staff, and the transfer of information from one to another in a working atmosphere.

2. This Paper deals with the specific field of thermal power stations, but the principles are applicable in other fields. A typical arrangement which might exist in a commissioning phase of a thermal power station is shown in Fig. 1. It pre-supposes a substantial previous experience with similar plant on the part of the owner and the availability to him of suitably qualified and experienced staff. In the UK for example, the owner invariably drafts into a new plant operating staff who have been trained and have gained experience in the owner's existing plant. He will have an on-going recruitment and training programme to maintain an adequate staff supply on his complete system and, as a result, it is easy to man a new plant.

3. In the case of an installation in a developing country where the local experience of power station operation may be small, the objectives are no different. The owner must end up with the plant being operated and maintained by his own staff in a competent

*Manager, GEC Turbine Generators Limited, Rugby

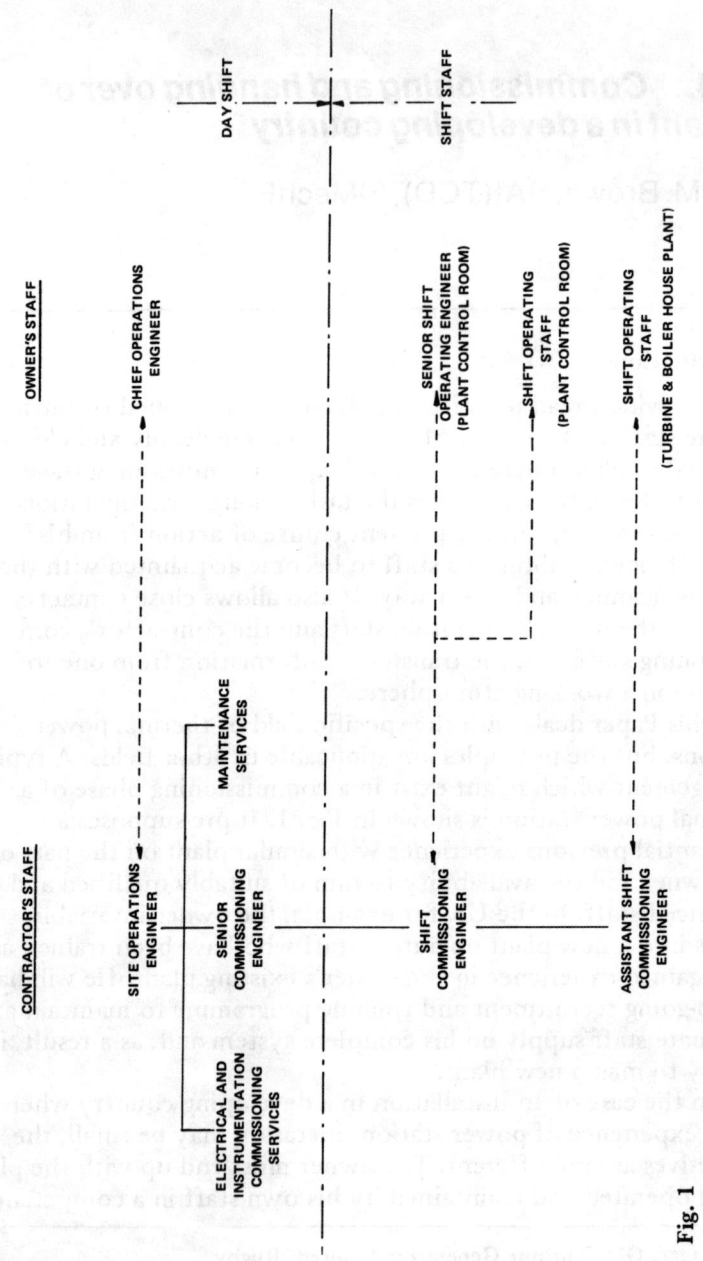

Fig. 1

and efficient manner. A body of trained personnel has to be produced for this purpose. In the long term such staff must be local nationals rather than expatriates. The purpose of this Paper is to examine the various steps that can be taken to produce a body of competent and efficient staff, even though an extensive on-going recruitment and training programme may not have existed in the territory concerned.

Areas of work

4. There are four broad areas of work which must be reviewed. These are set out in Fig. 2 and opposite each is an indication of who is responsible for carrying them out. For the purpose of this Paper the chart has been drawn up on the basis that the contract for the plant has been let on a supply, installation and commissioning basis.

Commissioning of individual sub-systems

5. The commissioning of a thermal power station does not take place in the form of a single operation. Many of the various subsystems can be set to work separately and handed over to the owner for operation prior to the plant as a whole being set to work. Two simple examples are the station cooling water system and the turbine generator lubricating oil system. The process of pre-commissioning and commissioning is an item-by-item, system-by-system one. The responsibility for the process lies firmly with the contractor, but there is no doubt that close involvement by the owner at this stage is one of the most productive 'education' areas for his staff. If this involvement can be developed and a relationship with contractors' commissioning staff established, then the owner's staff can be engaged in actually 'doing' things, rather than only 'monitoring' or 'checking'. In these circumstances the taking over and operation by the owner of completed subsystems can be done with confidence. The process can be continued until a large extent of the plant is covered.

AREA OF WORK	RESPONSIBILITY
COMMISSIONING OF INDIVIDUAL SUB-SYSTEMS	CONTRACTOR
COMMISSIONING OF TOTAL PLANT AND RAISING POWER	CONTRACTOR/OWNER
LONG TERM OPERATION	OWNER
LONG TERM MAINTENANCE	OWNER

Fig. 2

Commissioning of total plant and raising power

6. Central power stations present unique problems at this point. Before power can be raised the generating unit must be synchronized and connected to whatever grid system is in use, be it large or small. This operation and the subsequent one of controlling the rising and lowering of power must be reserved to the owner's staff since it involves co-ordination with plant and systems outside the plant which is being commissioned. Thus, this second area of work must be designated a joint contractor/owner task.

7. The interpretation and analysis of measured data relating to plant behaviour, especially when abnormalities occur during initial operation, are very much the province of the contractor and his commissioning staff. Corresponding problems of operating the plant in relation to the external grid system and its need and peculiarities can only be resolved by the owner's staff. A high degree of co-operation and working together is called for. This will be much more readily achieved if contractor/owner relationships have been well developed in the pre-commissioning phase.

Long term operation

8. This area of work is an owner responsibility, and a large professional staff is needed who will not only operate the plant in both normal and abnormal circumstances, but will also monitor its performance and make all necessary on-going adjustments. Key areas of knowledge are the plant itself, the operating characteristics of the various components of the plant, the system to which the plant is connected, procedures to be followed when abnormalities occur, etc. In addition there is a large specialized administrative system to be organized and kept going. Owner involvement in the two earlier areas of work will make a substantial contribution to success in this one, but will not be sufficient by itself. Previous experience on the part of key personnel in the day-by-day operation of similar plant is almost certainly necessary if reasonable performance and availability are to be achieved.

Long term maintenance

9. Again, this is an owner responsibility and covers a wide

variety of subjects varying from the mundane, such as good housekeeping, through to the sophisticated, such as the planning of routine overhauls and the execution of such work on high technology plant. In the long term it is often the quality of work in this area which determines the plant availability.

10. The owner must provide himself with staff who are sufficiently well trained to maintain the plant on a daily basis and to plan major overhauls in an orderly manner. However, it is open to debate whether he need carry staff capable of carrying out a major overhaul. Apart from questions of high technology and of the need for close attention to detail, there is the problem of doing work at sufficient speed. In the industrialized countries, plant owners often engage plant contractors' service organizations to get the benefit of speed, which in turn can only come from a repeated execution of similar work. The same argument applies in a developing country.

Training of the owner's operating and maintenance staff

11. One of the best options for the owner of a central power station is to operate and maintain it with his own employees rather than depend on services provided by plant suppliers or other agencies. On the assumption that there is available to the owner a supply of the various grades of staff, e.g. engineer, technician, supervisor, skilled artisans, etc., with a reasonable background of theoretical and practical technical training, and with some element of experience, however small, of the type of plant in question, in this section some of the steps that can be taken to train this staff to an appropriate level will be considered.

Training in the plant contractor's headquarters

12. Key personnel from the owner's organization can benefit from attachment to the plant contractor's headquarters during the design and manufacturing phases. The personnel must be those designated to be in charge of the plant during its early operation, and must be additional to those attached for progress, quality and other similar purposes. Inevitably, the people so attached are in an

'observing' role, and it is difficult for the contractor to give them a 'doing' role in any meaningful way. However, the process can be a constructive one if both sides are convinced of the need for information and knowledge to be transferred, and are committed to successful plant operation.

13. A formal training programme of lectures, demonstrations, etc., based upon design concepts, drawings, erection and commissioning instructions, and operations and maintenance manuals, can be conducted for attached personnel. Such a programme provides a solid base from which the owner's staff can work and enables them to get more from their period of attachment.

Training with similar plant being installed or operated elsewhere

14. This again is a training method for key personnel on the owner's staff. If the contractor is installing and commissioning a similar plant, they can join his site staff and be exposed to situations identical to those which will occur in their own plant. Opportunities for doing as opposed to observing are substantial.

15. On the purely operational side, and quite apart from installation and commissioning, many owners and utilities in the industrial countries offer training programmes to utilities in developing countries. It is here that an in-depth understanding of daily operational situations can be increased, covering both the normal and the abnormal. The level of success achieved depends upon two factors: the length of the training period, and the degree to which the trainee can become involved in the work and given responsibility. Experience gained by operating staff depends upon the number of different situations which occur during their training. The longer their training, the greater will be the variety and extent of these situations. The one week visit to a power station is unlikely to add much to operating knowledge, a six month spell will do substantially more, and clearly the longer the period, the greater the benefit. In the end cost/benefit considerations will settle the matter.

16. The question of performing the functions of an operator in a power station owned by somebody else, poses problems. No owner of an efficiently run plant will be enthusiastic about allowing people over whom he has only indirect control, and

whom he did not select, to take decisions and to press buttons.
The same problems arise in the case of outsiders working on the
overhaul of technically advanced machinery.

Training during the course of installation and commissioning

17. Training opportunities are at a maximum during the installation and commissioning of plant at site. The actual hardware,
the owner's staff directly associated with the plant, and the
contractor's staff allocated to the plant are all available. These
factors and many others, if controlled and harnessed, can make a
substantial contribution to the effective training of the owner's
staff, to a straightforward hand over of the plant and to its long
term efficient operation.

18. From the start of installation, steps must be taken to involve
the appropriate owner's staff at working level. This is a progressive
matter; the numbers must be built up gradually. The aim at all
times must be true involvement rather than something which is
only observation or checking. Success in achieving this aim
requires a clear statement of intent and a clear definition of the
roles and responsibilities of both parties. However, it also depends
to a great extent on the personalities of the people concerned and
both contractor and owner need to recognize this and select their
personnel accordingly.

19. A formal training programme similar to that in paragraph 13
can be used on site with even greater advantage. It can be extended
to cover all grades of the owner's staff and can be related to the
actual installation and commissioning work going on at any
particular time. To be effective such a training programme must
be designed to do more than describe the whole scope of the plant
being supplied and its various operating modes. It must make use
of all the modern educational processes now available, so that
trainees have the best possible opportunity to advance their
knowledge.

Training in plant maintenance

20. The two separate aspects of plant maintenance have already
been referred to: the tidy and orderly conduct of day-by-day
maintenance affairs, and the planning organization and execution

of major overhauls. For the first of these, training is best given by example. Good housekeeping starts at the installation and commissioning phase, and no better opportunity will ever occur for setting good standards. The greatest contribution that a contractor can make in this area is the insistence upon and achieving of these good standards.

21. The essential element in training for regular major maintenance is the actual carrying out of the work by owner's staff under contractor supervision. Obviously opportunities for this are minimal during the installation and commissioning phase. Hence, there is no alternative to a continuing contractor involvement on a decreasing scale in routine major maintenance of plant over the first years of operation. As time passes more sections of the plant can be dealt with completely by the owner's staff.

Means of additional support

22. During a period when an owner's staff is being trained, there are a number of means of additional support available to the owner.

During commissioning and initial operation

23. On the assumption that commissioning is a contractor responsibility the owner's need in this period is for trained operating staff. There are a number of organizations and utilities throughout the industrialized countries that offer the services of a team of people to deal with this situation. Such a team can be totally responsible for plant operation and include in it all grades of staff. Alternatively, they can confine themselves to a supervisory role. Another possibility is for such a team to be used on an 'alongside' basis whereby various members of the owner's staff are matched by opposite numbers in the temporary team who will provide the necessary guidance and advice.

For maintenance purposes

24. For actual conduct of major plant maintenance, contractors' teams are widely available. Such teams can normally be tailored to

an owner's exact requirements whether taking complete responsibility at all levels or providing supervision only. Assistance is available from various utilities in the planning organization of long term maintenance.

Conclusion

25. The handing over and long term operation of plant in a developing country is often seen as a major problem. More often than not the problem stems from failure to examine the situation sufficiently far in advance and failure to make suitable plans. If this is done the difficulties which occur can be confined to levels no greater than those experienced in the industrialized countries.

Acknowledgement

26. The author acknowledges the help of his company, GEC Turbine Generators Limited, in the production of this Paper.

Discussion on Papers 11-14

Mr Thomas

1. Paper 13 deals with broad aspects of managing a contract in a developing country, and I would like to bring these into focus by referencing them to a particular project — a large multi-disciplinary project carried out in Eastern Europe by a British consortium known as the Sadova Corabia Project. The consortium was led by Taylor Woodrow International who carried out the design, supply of construction equipment, and some material and supervision of the works. Members also included GEC, Sigmund Pulsometer Pumps and Vickers who were responsible for the mechanical, electrical and irrigation components of the project.

2. The contract took approximately eighteen months to negotiate, and although the technical specifications were very detailed, the requirement of the fixed price contract was to provide a minimum deposition rate of water from sprinklers on an area of 300 square miles.

3. The cost of preparation of our offer in 1969 was about £100 000 and it would probably be three times that amount today. It is interesting to note that the Bosporus Bridge contract was signed at about the same time, and it is difficult to see this fixed price contract situation being achievable today.

4. The contract was financed by ECGD backed UK finance and another feature of it which is common to developing countries was the existence of a parallel counter purchase agreement for agricultural produce.

Mr Brown

5. Operation of plant after installation is a skill in its own right. If plant is to be handed over and operated effectively, then this fact must be recognized well in advance and appropriate plans laid. Only rarely do contractors who supply, install and commission plant, have the necessary skills for its long-term operation. It is within the owner's organization that these skills must be available, and if they are not, arrangements must be made to develop and provide them.

6. There are some inherent barriers to the handing over of plant. For example, an owner's organization which has not previously operated sophisticated equipment will have a natural human fear of operating for the first time, and this creates a barrier. Again, the contractual arrangements necessary create barriers. As completion approaches, a contractor presses vigorously for the warranty period to begin while an owner naturally resists this pressure, to satisfy himself that the plant is in fact complete.

7. Paper 14 was based on this background information. Its object is to foster the view that plant operation is a particular skill which must be planned for by both owner and contractor from the beginning of the project.

Mr J.D.T. Kirk *(Marketing Director, Kier Ltd)*

8. The time available for the preparation of design and bids for North Sea platforms in most cases has been too short. In one case Kier had five weeks to prepare a design from scratch and prepare the bid. This involved designing a massive structure, with little previous experience of similar work, with many uncertainties and scanty information. It involved finding a location to build the structure, getting planning permission and some quality assurance from the certifying authority, insurance and so on. In such circumstances competitive tendering cannot have much meaning.

9. The amount of effort involved in the European construction industry in connection with the North Sea has been considerable and largely abortive. Large sums have been spent on research and development and on setting up some base facilities which will

never be used, and the oil industry generally came very close to being irresponsible in the way they encouraged people to jump into this field.

10. Apart from the need to allow more time for design, what contractual arrangements could be made in the future?

11. Much criticism is made of the poor completion time on UK construction projects, but there is rarely any differentiation made between civil engineering work and other sectors. In my experience it is rarely civil engineering work that is at fault, even though it is used mostly as the buffer operation. Work involving only civil engineers, such as motorways, produces much better results.

Mr H.S.G. Knox *(Director and Chief Engineer, Cleveland Bridge and Engineering Co. Ltd)*

12. The client company of Shell UK Exploration and Production planned originally to have the Brent 'C' platform and Cormorant 'A' built in a period of 2½ years, with apparently 1½-2 years being the net construction time for each. In fact both structures were built almost simultaneously but took a total period of 3½-4 years to construct and were not completed until 1977. Would it have been better to have planned to build the two in succession thereby lessening the pressure on resources? This might have enabled completion of at least one in 1976, or even one in 1975 and the second in 1976.

13. The cost of the two structures is stated, excluding inflation, to have increased from about £57 million to £84 million. Presumably the client did not pay for the contractor's default and therefore most of this increase must have been the direct and indirect effect of variations. The design was not completed at the time of tender — something which should be avoided.

14. In contrast to the traditional British system in the field of major bridge work and heavy structural work, more time is often taken in the United States between the concept of a project and going out to tender, with the object of achieving a form of contract which has few or no provisions for re-measurement or variations. But the result under the North American system has often been that the construction period following tender has been shorter and completion has probably been earlier relative to the concept of the

213

project. A further result has certainly been a considerable reduction in the commercial administration of the contract, including an almost entire absence of our quantity surveying effort, and fewer contractual disputes. Even under the British system we have experienced great differences between the performance of contracts which have been well or imperfectly engineered before going out to contract.

15. The Bosporus bridge was not a project at the limit of knowledge, but it was subject to variations which were due to a premature start to contracting and the fact that the form of contract permitted such variations. The programme demanded was tight. In this case the effects were not to create labour market problems but to create equally adverse market positions for the contractor towards suppliers and shipping, and against local laws and practices.

16. The significance of cash flow can be much affected by the contract programme. If this is tight, by concentrating procurement or, say, the level of a large reimbursable element such as customs duties, relative to the payment intervals and terms of payment, it can be shown that the maximum funding required is increased quite disproportionately, and although this may apply over a shorter period, the total financing cost may also be greater. Experience in most countries appears to be that very tight programmes on projects requiring large scale resource use or being subject to change are in fact rarely achieved. Clients might benefit by accepting this. The cost of a particular construction programme can only be properly assessed by a contractor for his own circumstances and methods. Clients should perhaps invite offers for varying programmes including the tenderer's own assessment of the least cost period. The results might be found to be of advantage to him, including those for high revenue earning projects.

Mr M. Milne

17. Where a client is a government department and involved in large projects which can run through several administrations, it is constrained by policies such as the economic situation which may change between parties, or even within parties of different complexion. This can mean that even where a contractor might be

willing to carry on in a certain way in dealing with his labour force, government requirements dictate that he shall not behave in that way, even though the Department of State concerned might for its own purposes wish to see the contractor so behave. Taking the oil situation as an example, the important thing is that the oil must flow. Therefore, a contractor operating in that sort of environment will undoubtedly want to keep the job running as quickly as possible regardless, for example, of the pay policy of the government of the day.

18. To what extent does an operator like Shell, an international body with pressing demands for its products, and where the products are important, find the constraints of government policy operating against their interest as an organization?

19. In the UK we also have the operations of quasi-autonomous national government organizations, such as the Coal Board, railways and so on, which work directly within government policy and yet are outside Government. They are also constrained in the way in which they can operate.

20. To what extent did the Government of Turkey in their policies, laws and regulations influence the way in which the Bosporus Bridge contract was run, and to what extent was the profitability, or completion date influenced by government policy? What is the difference between operations in the UK and operations in a developed country such as Turkey?

Mr F.G. Johnson *(Chief Civil Engineer, North of Scotland Hydro-Electric Board)*

21. The Brent 'C' platform and Cormorant 'A' were complex, virtually prototype structures. Would there have been advantage in putting the design in the hands of a consultant, particularly as there was a considerable period before going out to tender? If this was not considered then, would it be considered for future projects?

22. Was the fixed price contract adopted the most appropriate type of contract? Would a cost reimbursable contract with a target cost and probably bonus/penalty clauses for both performance and cost have been more appropriate?

23. If parts of the structures were fairly simple and straightforward, would it have been worth inviting tenders for these

sections on a fixed price basis, thereby leaving the more difficult sections, which had not been finalized in the design, to be covered on a cost plus basis?

24. What sort of contract would you consider if you were building them now or in the future, in the light of your experience?

Mr P.A. Cox *(Partner, Rendel, Palmer and Tritton)*

25. On the question of inter-union problems, whilst the management/labour relationship is talked about as being the important one leading to a successful conclusion of a project, many of the problems stem from the differences of opinion within the labour part of the organization, and inter-union problems are obviously one of these aspects.

Dr Klein

26. Something like building a bridge over the Bosporus is so exciting and romantic to the lay person that it must carry motivations of its own. In less exciting situations, people start putting in stimuli of their own, and sitting around in Tooley Street may be a situation of that kind.

27. Essentially, one is talking about the management of a temporary system, and my suggestion would be to put the question in those terms to the parties at the beginning. The question then becomes: How do the parties involved, including the different unions, preserve their individual continuity, their individual interests and their relationship with members, while at the same time co-operating in a temporary system? Is the problem ever put in those terms? The site procedure agreement in particular is one way of doing this but are there any more ways?

Mr G.W. Skinn *(Manager, Engineering Services, Teesside Division, British Steel Corporation)*

28. As a nation the British have been unable to meet their plans. How can the contracting industry, within the framework of the

competitive situation, get together and plan its overall resource requirements against a forward work-load which keeps changing so often?

Mr Cox

29. Perhaps some of the problems in the construction industry stem from the fact that the management has always been in fairly close contact with labour, and that systems are now being imposed on the construction industry which have been found necessary in manufacturing industry, where contact between levels is not so good. There is a large area to be explored in the continuation of the relationship from management down to the labourers on the site.

Mr T.D. Kershaw *(Consultant)*

30. As the NEDO *Report on Large Sites* (reference 2 of Paper 10) indicated, one of the factors contributing to industrial unrest and consequent delays is the limited authority of plant manufacturers' site agents as compared with the authority granted to the agent of a civil engineering contractor. The reasons for this are understood but although various palliatives and solutions have been tried, a solution to the problem seems far away.

31. Could the specialist installation contractors, similar to those who operate on overseas projects installing any fabricator's plant, be established in the UK and provide the solution to this vexed problem?

Mr P.A. Thompson *(Project Management Group, UMIST)*

32. The engineering profession attracts individuals with a logical mind and their initial technical training can encourage them to identify problems and take reasoned decisions, thereby making good project managers. Few of the successful managers of my acquaintance have, however, received any subsequent training in management skills. In most cases they have developed their

natural talent through a progression of responsibility for jobs of ever increasing size and their considerable capability when dealing with people owes much to experience in the armed forces.

33. The position has greatly changed. There are few jobs of any size in this country and the young engineer is rarely eligible for responsibility overseas. With the demise of the small project communication between tradesman and engineer js becoming increasingly difficult. How then do we identify and develop the project managers of the future?

34. The few mid-career training courses currently offered by university departments and the Institutions of Mechanical and Chemical Engineers are mainly concerned with specific techniques whilst the general courses offered by business schools are remote from the problems of the project manager.

35. If we wish future generations of project managers to be drawn from the engineering profession we must give urgent consideration to this problem and the Institution of Civil Engineers should take a lead in formulating a training programme. It is my opinion that a cadre of potential project managers sponsored by different firms and organizations should be offered, as a group (or groups), regular periods of intensive training in both management theory and techniques and on those relatively few sites where the essential ingredients of practical management can be experienced.

Mr P.A. Banks *(Partner, John Taylor & Sons)*

36. In many cases the clients in developing countries, even though they are probably government ministries with bottomless purses, seek to load all risks on to the contractors because of their own lack of experience and ability in managing projects. They modify the FIDIC Conditions (reference 3, Paper 7) out of all recognition to ensure that this is the way the contract is finally drawn, and rely on their own rigid internal financial department regulations to ensure that anything not covered by the Conditions can be made the contractor's responsibility.

37. The private justification for this attitude is often given as the need to guard against the difficulties which would arise with local contractors if the Contract was written in any other fashion. The role of the traditional Majlis or court where those aggrieved

can approach the man with the ultimate power of decision as a last resort, can give some confidence to honourable contractors who seek to complete a satisfactory job according to contract, but incur heavy losses in so doing. Nevertheless the Middle Eastern approach has resulted in high prices to match the high risks.

38. The need for association with local organizations is often crucial to success, but this is difficult to arrange *ab initio* during a hurried pre-tender reconnaissance in a new country. Therefore, if large capital projects are to be sought, some years of familiarization may be necessary to assess and choose the right association long before the stage of putting together tenders is reached.

39. The use of English gives us an advantage but it is essential to have good communication between the man in the trench and his foreman, through perhaps a gang leader, if good productivity and good quality work is to be achieved. The best of our foremen, like our servicemen, can do this but some positive training or aids can often improve matters substantially. As an example, I have recently seen illiterate Yemeni labourers bottoming trenches with the aid of the red dot of a laser beam on the handles of their shovels.

40. It is often attractive financially for offshore contractors from an industrialized country such as the UK to provide management only to a local contractor engaged on a large project. This seems to accord with the wishes of the host government for the transfer of technology and the economic growth of their own contracting sector. It can, however, be disastrous unless experienced competent first-team staff are provided by the managing contractor, and the local contractor gives them financial control, even if this is within some ceiling figure and constrained by budget and cost control techniques.

41. The consulting engineer's job often includes the task of creating a client and an employer for the contractor after the contract is signed. Even though most major financial decisions are retained by the client with only advice sought from the consulting engineer, no mechanism may exist for obtaining these decisions during the course of a contract. Young inexperienced administrative or engineering staff in the client's departments have to be taught about variation orders and also to appreciate the contractor's need for cash during the course of a contract if work is to proceed. Mr Thomas suggests that in some circumstances we can be a further

burden to the contractor by seeking to ensure a more than adequate job. I would say that one thing the British consulting engineer has to sell in the market which other countries' consultants may not have, is the ability to see that a good job is achieved, and this is undoubtedly one of the reasons why they are retained. Furthermore, I believe that a good job is sought from all contractors, not just the British ones.

Mr A.R. Parish *(Deputy Chairman and Chief Executive, W.S. Atkins Group Consultants)*

42. It seems to be assumed that a project starts when someone decides to build something and ends with a completion certificate. This is certainly not true from the viewpoint of the potential owner. It is therefore a pity that no mention has been made of feasibility studies, and of the ability of computer programs to up-date such studies. Equally it is unfortunate that minimal attention has been given to the human requirements to enable industrial projects to reach the designed output, since failure in this way has caused more economically unsatisfactory projects than capital cost or time overruns.

43. Engineers tend to forget that the success of a project, both in the construction and operation phases, depends on people. This requires the development of basic manual skills as much as technical or managerial ability, and the manual skills may have to be imparted to individuals with no experience at all in an industrial society. Any failure or inability to develop such competence means the necessity of importing expatriate assistance until the indigenous level of skill and competence can be built up. However, this solution imports another problem — good 'do-ers' are only accidently good teachers and the expatriates have to achieve a transfer of technology within an unfamiliar environment.

44. It must be remembered that the people who are operating, maintaining and managing these projects do not just have their lives inside the power station gates or the works gates, but also have an outside life. Projects have been built which are technically competent but which provide no shopping facilities, poor general medical facilities, poor recreational facilities, and so on. It is perhaps regarded as a political problem, but it is also a practical

problem for the project. On one project the absence of this human infrastructure resulted in a 45% labour turnover in the first year of operation. This can totally disrupt a training programme and destroy the projected economic benefits of the project.

Mr R.J. Bridle *(Department of Transport)*

45. The consequences of large projects on the indigenous population seems to be a wider problem than merely commissioning. A large project, larger than the existing infrastructure, may cause economic disequilibrium. To what extent should engineers take this consequence into account in advising on a project?

Mr M.H.S. Muller *(Technical Director, W.S. Atkins & Partners)*

46. In many developing countries, such as Algeria, especially those which have adopted a socialist type of government, the client organizations are mostly governmental or quasi-governmental. This means they have, through their own structure, to take control and manage their projects in detail, unlike similar clients in developed countries who feel they have institutions in which they can entrust the responsibility. The staff of a consulting engineer becomes integrated with the client's organization. The consulting engineer has to provide much of the design, management, organization and co-ordination of the project for which the client does not necessarily have the resources. In doing this, it is obviously necessary to involve the client's own engineers all the time, who are often bright young men without much experience.

47. In Algeria foreign civil contractors are not allowed. In consequence, Algerian contracting companies have formed partnerships with European firms who provide them with staff, who then become integrated into an Algerian organization. This is not the same as a management contract: the control and final word lie with the indigenous partner.

48. Lack of infrastructure and resources in the country leave a project vulnerable to delays since shortages cannot be set right by quick imports. The client may try to act as a kind of general supply contractor by letting supply contracts in advance for those

items that he thinks most likely to run short – a difficult and complicated arrangement.

49. The organization, planning and thinking ahead for working in such countries has to be done down to a level which is taken for granted in many cases in the UK and in Europe. It must be simple and thorough, because it cannot usually rely on sophisticated techniques to support it.

50. This transfer of technology is a growing situation in developing countries, but is hard work.

Mr J.W. Rogers *(Assistant Managing Director, Taylor Woodrow Construction Ltd)*

51. My remarks are made in the wider context of multi-discipline projects rather than civil engineering alone. In my experience there is a dichotomy between the operational department of a client and his project group (being by definition those people who are responsible for the development of new plants and extensions, or refurbishing existing plants). The operational experience must be brought in at the beginning of the design and remain throughout the design and commissioning period.

52. The following is an example of a method of proceeding to establish concept recently used by a client. The client organized a brain storming session at his headquarters, lasting about one week. He gathered together representatives of all aspects of their work – research, operations, buying, project and so on. The client's Project Manager explained the major parameters of the plant to be built and set out every possible way of achieving these parameters, asking for comments as he proceeded. Having achieved a complete list, he reviewed each item to decide (with the help of the meeting) whether it should be finally included or not. The object of the exercise was to determine the brief for the feasibility study and to obtain the views and experience of all those who could sensibly contribute. One day was spent on each major section of the plant and working parties were set up to review certain aspects and to report back for review at appropriate intervals. At the end of five weeks, with one intermediate review, the brief was established with a milestone programme for its achievement. This was a new experience for me and I thought it extremely worthwhile.

Mr Milne

53. An engineer overseas, with a new piece of equipment, runs the risk, no matter how well constructed the plant is, that at some time he is going to be in a near-failure situation. To what extent does either the contractor or those who have employed him give the information to the operator of the kinds of things that can go wrong and prepare him to judge that situation and to recognize when he needs help and support of somebody outside his own organization? Failing that kind of communication, conveying of information of this kind cannot be done in a short period. In the transfer of technology to undeveloped or developing countries there is a corresponding transfer of technology from the developing countries to the developed countries. As this experience extends, the reverse flow will be as meaningful and as important as the flow in the other direction is perhaps at present.

Mr R.W. Postlethwaite *(Peter Fraenkel & Partners)*

54. Many overseas clients are not aware of the necessity to apportion risk properly between client and contractor, or they do not fully understand it. I have seen a graphic example of the considerable additional cost which can be incurred on a contract by writing in onerous conditions. In two neighbouring countries, two similar jobs were tendered for — one on straightforward FIDIC conditions (reference 3, Paper 7), and the other on a very much onerous form of what was originally FIDIC. The former tender was about three times less than the latter.

Mr J.T. Edwards *(Partner, Freeman Fox & Partners)*

55. With a small project, factory or site, the management can go and inspect the work and can size up the situation in a face-to-face meeting. In projects which are large geographically — where the project is executed on a small site but has design and manufacture and so on going on in different parts of the world — it is quite impossible for the top management to see literally what is going on. Development must be made in the ways in which one can set

about controlling and managing these projects where the top
management is extremely remote.

Dr Klein

56. It seems to me that there must be quite a difficult issue
about 'handing over the baby' and also about deciding whose
'baby' it actually is. In the same way, there must be a great
temptation to believe that local labour cannot cope, even with
training, and then to turn that into a self-fulfilling prophecy.

57. If training programmes do not succeed in getting local
labour up to performance, maybe it is the training programmes
and their design that are inappropriate.

Mr J.M. Holloway *(Deputy Project Manager, Thames Barrier,
Department of Public Health Engineering, GLC)*

58. Clients in developing countries frequently wish to have the
most up to date and advanced equipment on the market, but there
is a strong case for the provision of what has been termed 'appropriate technology'.

59. The scope for simplification obviously varies with the nature
of the installation but whenever possible the engineer is duty
bound to adopt designs appropriate to the real absorption capacity
of the client's nation. A paramount factor to be considered in
the design of installations and choice of plant is the operational
capability.

Mr Allcock

60. Mr Kirk made a point about the oil industry verging on the
irresponsible. I may be leaving myself wide open to trying to make
two wrongs into a right here. This is related to Mr Knox's question
as well. We think we did ourselves a disservice; we did not, however,
think we were verging on the irresponsible. Most of the contractors
who bid for our work had several weeks or months of preparatory
conversations and discussions before the hard tender documents

came out — abstract as they may have been. Five weeks sounds in extraordinarily short time, but I think there was an extensive lead-in, and the industry knew of the short tender period.

61. It was suggested that we would have been better advised to have given two contracts in succession to Ardyne Point. In a pre-tax situation of a production platform in the North Sea, early production is absolutely paramount. For example, a platform is going to produce 100 000 barrels of oil a day, and one can get that oil flowing a year or two earlier. In present-day terms, and recognizing the tax scene with oil at 14 dollars a barrel, there is no way in which we are going to wait two years to award a second contract. If it cannot be done at Ardyne Point, that job will be given to somebody else. It was not we who were being irresponsible — if that word can be used. Within a couple of months of McAlpine accepting our two contracts, they accepted a third contract, and built three platforms together at Ardyne Point. I will not comment on which side was right and which was wrong. McAlpine-Sea Tank were obviously confident that they could build three platforms at the same time within the contracts they had.

62. With respect to labour, Mr Kirk said that civil engineering contracts on the whole remain more on time than those of other trades — mechanicals or electricals possibly. The problem at Ardyne Point was not strictly a union versus management one; it escalated itself into being a conflict between two unions. It had little to do with the client or the contractor; two unions were in some sort of dispute which seems to have caused much of the delay.

63. With reference to Mr Milne's question (paragraph 11), we aim and will continue to aim, in terms of contracting work, to complete designs and get out on lump sum bidding. This is not very popular in the UK, but nevertheless it does represent a major form of cost control. The changing circumstances have always forced us, over the past few years in the UK at least, to respond when things, such as Government legislation, change. The counterpoint is the disinterest of the Government when major issues are being debated on site with respect to industrial relations and industrial disputes. One particular case at Ardyne Point was where the civil engineering industry agreed that the work-force at Ardyne Point would get an 'offshore bonus'. It was agreed that the men were working on a big job for the North Sea and therefore they were entitled to so many pence an hour by virtue of the fact that

this piece of work was going to float out to the North Sea. Why that made the work more costly, or why people had to be paid more to do exactly the same job, will remain a mystery.

64. The other point where there is growing concern is this question of severance payments at the end of a contract — this deliberate going slower which besets civil and mechanical contracts in this country, related to North Sea activity at least, leading up to the labour saying they will not finish the work unless they are given a thousand pounds, for example. The Government does not seem to want to get enmeshed in these disputes. When these major issues appear, it is left to the unions and their shop stewards to try and do the best they can in the circumstances, which normally finishes up with the client paying the money.

65. In reply to Mr Johnson's first question (paragraph 14), these massive concrete gravity structures go through several phases in their construction, principally related to their marine stability when they are floating inshore and construction is continuing, with the overriding consideration that the industry would wish to take advantage of the concrete structure and put as much deck load and all the process facilities to be used offshore on that structure before float-out or tow-out to site. This means that the design of each of these concrete structures has been specific to the site at which it was being built. It accounts to a degree for the significantly different forms of the various concrete structures manufactured in different countries by different contractors. Much of the contrast in the design of the one built in Europort at the mouth of the Maas, those built in the Norwegian fjords, two built by McAlpine-Sea Tank and the other one built by Howard Doris, was to do with the draught on which the structure had to float out. I have forgotten the exact draught for Ardyne Point, but it is limited by towing the structure along the north-west coast of Scotland. The Norwegian structures have been towed out at about 120 m from Stavanger, but from the offshore island in Norway, Stord, they are turning out at far greater draughts. It would have been pointless for us to appoint a consulting engineer unless we told him which contractor was going to get the job ahead of time, because the design had to be specific to the site at which the construction was to be carried out.

66. With respect to the form of contract that we would go for in the future, it is very interesting to hear the continuing debate

and discussion about the form of a contract. I think the UK contracting industry has become progressively more nervous about its ability to estimate how much anything is going to cost, largely because of labour unrest and the problem of when material is going to be delivered. If one has a form of contract, where one is not confident that things are going to happen exactly right, then obviously one must go for some kind of reimbursable target with a fixed fee bonus in that style of contract.

67. In the more stable conditions which apply in Europe and the USA — although they have their strike problems as well — we have successfully built two concrete structures. One such structure in Norway was delivered on time without any harangues or disputes at the end of the day, virtually for the sort of money they said they were going to charge us, almost for the lump sum. There was discussion about variation orders and changes, and negotiation took place, but it was the order of about 10% of the contract value of that order. The UK, therefore, can argue one way. Other people are arguing in a different way; they prefer to have lump sums because there is a greater incentive.

68. Mr Skinn's question is interesting, because when this North Sea oil project started, a number of people were carried away pushing UK industry into this tremendous new area of activity. It was not confined to the creation of sites at which to build the structures; it was also to do with the process facilities, and various scenarios were drawn, notably by the Government Departments, illustrating continuity of work, stretching into the 1990s and even further, to a degree that most of the sites at which these structures, particularly the process facilities, were going to be built, were put together on the fundamental thesis that the work was going to have continuity.

69. It was impossible to bring the structure to the pools of work-force, so the work-forces had to face up to going to the sites where the structures could be built. In the case of the process facilities the opposite occurred. The facilities manufacturers, assemblers and fabricators were able to set up their yards in the centres of traditional industry in the UK — Glasgow, Newcastle, Teesside, Firth of Forth and so on. It was this continuity aspect which was introduced into nearly all the site agreements, almost to the extent that sermons were preached about this being continuous work and asking the work-forces to come to the sites. Of course,

within a couple of years, for many and varied reasons, not excluding the introduction of a major disincentive like the petroleum revenue tax, the oil industry's attitude changed, and its attitudes still hold many uncertainties. But the work has turned itself back into the old-fashioned construction site activity. It is all peaks and troughs and there is no continuity. I think that in the oil industry we tried to tell people that this continuity would never happen. However, it did not affect the philosophy that was passed around, and as a result most of the selling was done on a premise which was proved to be erroneous after two or three years; now, of course, a number of these sites have had to close, and a couple of them were never even used.

70. With reference to Mr Cox's point (paragraph 22), it is unfortunate that maybe the finest example of this 'togetherness' attitude was the site of the Norwegian contractors in the Stavanger Fjord. Now that the Norwegians are independently wealthy, and nobody can afford to go to Norway, one feels that the labour forces there are also beginning to react, not exactly in the same way as the labour forces in the UK but they are moving in that direction; this may well be because of the imposition by the Government of statutory wage increases. I have forgotten what the figure was — 1½ per cent perhaps — and the work-forces have rather taken objection because prices have risen so dramatically in Norway, as have taxes. The magnificent attitude that the Norwegians had three or four years ago, where management and work-force alike were coming to work in similar cars, sleeping in similar bunk houses, eating in the same canteen, coming to agreements a couple of weeks ahead about what bonus was going to be paid on specific jobs, and also where they carried on working in half a gale and pouring rain, were looked after with respect to welfare, and were provided with showers, changing rooms and so on, is probably more problematic in Norway now because of the effect of outside pressures.

71. Traditionally, in this country, there has been an identity in the road, railway and bridge building activities because of the need to be together in the mud on site. But with the setting up of sites for North Sea construction there is a move towards an industrial scene — a factory-type agreement on the site — and this seems to generate a distinction between management and labour. It is not all a one-sided affair. I think the work-force is also anxious

to promote this difference as being a good way of running negotiations.

Mr Dixon

72. Mr Knox refers to the normal American practice by which invitations to tender are not issued until the design and the documentation are at such a stage that there need be few provisions for remeasurement or variations, and by which contract administration and claims are reduced to a minimum. I am of the opinion that this is generally correct and that both time and money are saved.

73. He also refers to the large variations in cost which occur between a contract being executed in the minimun time or the economic time. In the case of Bosporus Bridge the entire cost was paid off by the collection of tolls in 2¼ years and the tight programme was well justified. There are many cases however where a contractor's submission of a lower price for a different contract programme might well be of advantage to the client.

74. With regard to Mr Milne's questions, I believe that the Bosporus Bridge could hardly provide a bigger contrast to the situations which he quotes. The bridge was demonstrably necessary as shown by queues of lorries several miles long waiting to cross. The Government policy was to have the Bridge complete at the earliest possible date. The precise date by which they wanted it was 30 September, 1973, because that was the 50th anniversary of the Turkish Republic; therefore the policy was consistent throughout.

75. Concerning labour, we negotiated a bargaining contract with one union to which every man joining the work had to belong. The agreement was honoured and there were no strikes or disputes. The only difficulties were in keeping to the termination clauses of the legislation, which were extremely strict, at the end of the contract.

76. I consider Dr Klein is perfectly correct about building a large bridge. To the engineers anyway it is fascinating and that is an enormous help in getting the enthusiasm that is required. The majority of those employed on the Bosporus Bridge were in the job for some three years; the problem of maintaining individual rights in a temporary system was not, in this case, a serious one.

77. Mr Cox refers to communication between management and labour and vice versa. Normally, in any company, or on any major site there is a hierarchy, where there is the manager who has his next level of authority, then the foreman, then the operatives — a natural method by which communication may be made. Similarly on the union side there are the union members, shop stewards and district delegates and again there is a hierarchy. Latterly, systems have occurred where these hierarchies are ignored. The operatives do not accept the decisions of their shop stewards and they in turn those of their district delegates. The management, because they are beset at the wrong level by the unions, do not work down through their foremen. Until there is some form put back into the system this is one of the great difficulties which prevents us having stability.

78. Sponsors must not only let the contract which is most advantageous to them but must also be seen to be fair in doing so. It is this latter factor which has led to the insistence on 'clean' bids. Sophistication in the manner of issuing contract documents and awarding contracts is leading to systems by which complete adherence to certain portions of the inquiry documents is mandatory but where there is room for qualification in other areas. Such systems must surely be to the benefit of both sponsor and contractor.

79. Mr Thompson identifies a very real problem. The facts are that an essential part of learning to be a project manager on a large project is through experience on smaller projects and that there is now a preponderance of larger projects. I agree that the answer lies partly with an initiative by the Institution of Civil Engineers in forming training programmes as suggested, but it is equally necessary that contractors operating major projects should organize them so that those in junior positions have the authority to take reasoned decisions. The tendency to centralized authority without delegation has many disadvantages.

Mr Thomas

80. I am sure that all of us on a big project have a feeling of paternity towards it. We see the thing handed over. We see it being

operated often far below its efficiency, and it is only natural that we should express our disappointment.

81. Dr Klein may not be aware that a developing country always wants to buy the most advanced technology. The idea that 'small is good' does not matter. When you get a Ministry with the wind under their tail, they tend to want the very latest, and this causes many problems. The other thing about operating staff is that the engineering background, which every child in this country gains through its toys and environment, is taken for granted.

82. I feel unhappy about Britain's participation in large projects because of the insidious effect of its high internal inflation rate. You may remember that the last Prime Minister signed a thousand million pound deal with Russia for large projects. Little of this has been taken up because we cannot offer a firm price.

83. We have a considerable problem in the large capital projects which also hinders investment, as people must know when they raise funds what the end cost is going to be.

84. Mr Banks mentioned the appeal to the Majlis and I agree that this is a final recourse for dispute. However, when this is looked at back in London by one's legal department in the Board Room, it does not quite give the feeling of security which British firms like to try to have compared with normal arbitration procedures.

85. Language and communication is certainly important. In Romania all our people learned Romanian, and this was a vital factor.

86. Mr Banks' next point covered management contracts. Everyone is now rushing into management contracts in an effort to get out of the risks involved in our inflationary position and fixed price contracts. You let them take the risk and we provide the brains, fine, but you must remember that you are selling the seed corn, and if we do this on a large scale and lose our red-blooded contracting ability British overseas construction will just be a memory.

87. With regard to consultants, Mr Banks skirted round the point I made most delicately which is that when we are making the comparison with British contractors we are not talking of the Germans, the French or the Norwegians; we are talking of people from the Levant, people from Eastern Europe. There is no doubt that in the end the pressure on a resident engineer to get the job done induces him to accept a level of workmanship which can be

less than excellent. I think this is unfair, because competitive contracts do not seem to be weighted to take account of the quality of work done by individual firms.

Mr Brown

88. Mr Parish's comments are much in line with my own thoughts. One point which he dwelt on was the question of our ability from a developed country like ours to train people in a developing country who have not had previous experience. We must admit that the problem has not been solved. It has not even been solved well in connection with the people who do the same jobs in this country. Bodies such as the CEGB rely largely for operating staff on people who are trained over a great length of time and whose experience is built up over the years in low level plants, then in medium level plants, then in higher level plants, with increasing responsibility all the time. That process is just not possible if you want to take somebody from a developing country who has not previously run a large power station and, in six or even twelve months, bring that person to an adequate state of competence.

89. The contractor suffers in the end if he cannot hand over the plant. When it is not operating properly, it is difficult to demonstrate that it is faulty operation and not faulty supply that leads to that situation. It is, therefore, in the contractor's own interests to find some way of solving this training problem, and I do not believe it has yet been found.

90. I think the 'we and them' argument is greatly overdone. We preach to all our people potentially going overseas about things to avoid, and the 'we and them' argument is at the top of the list. However, it is nothing more than a way of describing two groups of people. There is nothing wrong with it as a technical description, but to certain people in certain circumstances it is insulting, and that is the reason we do not encourage its use.

91. On the question of clutching the baby and not letting it go, a true story will illustrate the problem. We had commissioned a power station in a certain country and had long passed the date by which we should have left the site. Costs were still going up when in theory we had finished. Finally I instructed the site manager that he and his staff should leave. The reason he had not done so

earlier was the anxiety and nervousness of the client about operating this plant, despite all the steps that had been taken in training. Eventually, a day and time for our staff to quit the control room were agreed. Our site manager attended to represent the company on this formal occasion. What actually appeared were two soldiers. One of them held the site manager's arm behind his back, while the other one put a large padlock on the door from the outside. The site manager was informed that he was not to leave until he had given his undertaking that he would not announce any more threats but would stay in the country until he was given formal permission to depart. Thus, we have a great desire to let the baby go, but it is not always possible.

92. The central problem is about dealings with people. The technology and the engineering can be taken for granted, but we have not learned how to finish off our jobs properly in the human sense — in the making of a relationship with the client. This is true particularly with industrial projects where things have to operate, as distinct from static things like dams and bridges and so on. We do not seem to have come to an understanding about how to bring things to a conclusion in good working order, leaving everybody in a state of maximum happiness. Nobody has come up with a solution in this discussion; I have not come up with a solution in what I have written in my Paper. This does not arise from any unwillingness. Perhaps we are not motivated strongly enough to search for the answer. Maybe the contractual people need to find some better way of motivating us to do it.

Closing address

Sir William Atkins, CBE, BSc(Eng), FICE, FIE, FIStructE*

In dealing with large capital projects over the years, knowledge has been gained in the selection of people who make up total project teams — engineers, manufacturers, contractors and others — and in harnessing their great skills to the needs of the client and hence to the project itself, in a total overall strategy of execution. However, looking ahead to the next decade one can reasonably anticipate projects so large that they will dwarf the so-called jumbo projects. The rate of capital expenditure will be limited only by the availability of finance, materials and manpower, and we are increasingly expert at lifting those limits. The time scales can well extend beyond the end of this century.

What pressures are likely to generate these projects? There is a growing awareness that what in the past were considered as separate self-contained development activities — industry, housing, transport, health, education, water, electricity, leisure — are now seen as interacting elements of a single development plan. They must move forward together in a balanced complementary programme. Conflicting demands on scarce resources must be reconciled. They must be phased so that the first plan is viable in itself and, as well as meeting immediate needs, provides a good spring board for further development. Education, health and leisure facilities, according to the required standards, are essential.

Developing countries have the political will to embark on these major developments. They wish to develop whole regions, changing nomadic populations into agricultural or industrial communities, set up transport and communication systems to act as cohesive elements to unite their country, and feed, house, educate and

---

* Chairman, W.S. Atkins Group Limited

Management of large capital projects. ICE, London, 1978, 235–239

235

improve the health of whole populations. Once started, the whole project must be completed in the total viable whole. Programmes are set in years, rather than decades.

The developing nation can deploy and absorb the resources necessary to implement their plans. The commerical world can capitalize their natural resources of oil, minerals, fertilizers and so on. The international banking world can mount the funding operations to assemble the necessary finance packages.

The planning concepts and analytical techniques are now available to structure these massive developments. Computers can be programmed to handle the vast volume of interactive data to assess risk and optimize options. The planners have recognized and can handle the multi-disciplined interactive design process.

Individual parts of a giant project, jumbo in themselves, may not present great management difficulties to those who are used to them, but to aggregate them into one whole giant project, inter-relating the parts, would present a new generation of management challenges. Size in itself creates severe problems, demanding a strategy of direction and control that will span the world in its use of men, materials and equipment, and that is able to meet problems and conditions that may drastically change during the life span of the execution of the project.

It has been said that engineering and management are insepar-able. I would agree if engineering is to remain a profession.

Why do we criticize ourselves on human relations and then, according to what we all say about ourselves, do nothing about it? What really is the tool of motivation in human desires?

I have referred to giant as opposed to jumbo projects, and the significance of the scale of the project, the growth in size and com-plexity, and the additional dimensions of giant projects — logistical problems, the assessment and control of risks, the motives, objec-tives and priorities of the client which created his forward plan, and so on. The normal reason for building a major project lies in the demand for the product which can be priced at a level showing a commercially satisfactory return on the capital invested. Such a motive is clear. While this would be correct for a privately funded operation, most major projects, and all giant projects, are Government-controlled, and Governments may have quite different, but quite proper, motives. These different motives, essentially non-financial, may result in a different project from that which would attract an entrepreneur. It is, therefore, important to identify the client objectives and to define the manner by which they may be expressed throughout the project organization. The tasks of satisfying these objectives are already showing themselves in jumbo projects nearing completion – projects which have had no top management team which complements, or indeed fully sustains, the client's own organization. Without a competent client there will be no orderliness and no ultimate voice of authority. Therefore, creating the client, or completing his organization with those supporting skills that it lacks, is the first task. This task is necessary since it creates the only body that will, or should, see and understand the whole.

In the development of lesser developed countries, in the transfer of immense wealth from one group of countries to another, in the growth of unemployment in the industrialized countries, and in the reassessment of political thought and ideals, can be seen the birth of giant projects, and one can begin to realize what these events entail. There is a new, and perhaps exciting, generation of challenges, related to scales not yet fully encountered, but with which we are becoming so aware on the near completion of present jumbo projects.

A transient labour and operating force numbering tens of thousands, or maybe hundreds of thousands, will migrate into the development areas. They will need to be housed, fed and entertained. There will be clashes of social attitudes as different cultures merge. This has been seen to a small degree in Britain during the development of the North Sea oil field. In the early days of railway development, the gangs of navvies moving in contained communities from town to town presented similar problems. The phenomenon is well known. In Saudi Arabia there are large numbers of Americans, Britons, Iraqis, Pakistanis, Koreans and Yemenis, each with their own culture and their own language, living in their own communities, and participating in that country's major development programme.

Materials and components will need to be assembled from all over the world to remote locations. Designs will be done in one place, materials supplied from another, fabrication will take place somewhere else. This will present major problems in logistics, demanding the full resources of modern materials handling, and information handling, techniques.

Developing countries will wish to use to the full their indigenous resources of labour and materials. Design and construction methods will have to be tailored to meet this requirement. Again, scale will present new challenges. It may be necessary to develop complete new design standards and codes of practice. The training of several hundred construction workers is one thing. To create a construction industry with its supporting education and vocational facilities is a problem of a different order.

What is an exercise in cost control on a more modestly scaled project becomes a major item in a national budget. A slippage in programme could cause a crisis in a country's balance of payments. Product exchange — barter deals — will be common. Worldwide inflation and the movement of currencies become of major importance. Cost control must become international funding.

When the organization of a project is structured, it should be greatly influenced by the client's view, as this view uniquely identifies with the client's true objectives: to offer a project complete and fully operational as quickly and economically as is reasonably possible since the supply of the world's wealth available for such ventures is limited, and today grows slowly.

The project team, headed by the project manager, should have a clearly defined separate identity. Its members should, at the beginning, have a single undivided commitment to the project. The team will need to comprise a wide range of skills and hence is likely to be drawn from a wide range of services — the client's staff, operating advisers, managing contractors, quantity surveyors, countries and consultants. They must be dedicated in partnership to the project with a direct reporting relationship to the project manager. All the activities which directly influence the successful realization of the project — design standards, operating performance, contractual procedures, funding arrangements, and so on — interact with one another, and therefore should be under the day-to-day unified control of the management team.

Major projects are shaped in response to a complex set of political, social, economic and commercial forces. Inevitably during the course of the project the environment in which the project exists will change and patterns will build up which will seek to change the project; and indeed, where it is appropriate, and practicable, for changes to take place, the project should adapt. However, I believe that the project team should be protected from the pressures of the changing environment. They should never need to take their corporate eye off the project to look over their corporate shoulder. A project directorate should be set up on which all interested parties can be represented or make their representations. This directorate will act as a buffer between the project team and the outside environment. It should give clear policy directives, make the necessary resources available, monitor progress, and redirect as necessary. Protected in this way, the project manager and his team can get on with the project in hand, and can be selected simply for their skills of project realization without needing to be politicians as well.

As I have described it, the management of each project is a complex multi-disciplined activity, and it follows that the project manager would need to be a man of considerable stature. Will he come from the ranks of the professional managers from the business world — accountants, economists, the armed forces, the business school graduates, the engineers? Looking back into the history of major civil projects, there are revealed the names of Telford, McAdam, Brunel, and others, as earlier project managers. They dealt with the conception, parliamentary bills, finance, engineering execution and, in some cases, operations management. These engineers of the past conceived and managed projects engaging capital expenditure comparable in those times of technology and experience with those we are likely to be faced with in the near future. Each was a civil engineer, entering a profession because he wanted to construct those things which the civil community desired, and in which his experience made him a pragmatist, which experience we inherit.

It is interesting that most of our best project directors started their careers as civil engineers.

Printed in Great Britain by The Burlington Press (Cambridge) Ltd. Foxton, Royston, Herts.